DAMMING THE COLORADO

NUMBER THIRTY-FIVE
*The Centennial Series
of the Association of Former Students,
Texas A&M University*

DAMMING THE COLORADO

THE RISE OF
THE LOWER COLORADO RIVER AUTHORITY
1933-1939

By John A. Adams, Jr.

TEXAS A&M UNIVERSITY PRESS
COLLEGE STATION

Copyright © 1990 by John A. Adams, Jr.
Manufactured in the United States of America
All rights reserved
First Edition

The paper used in this book meets the minimum requirements of the American National Standard for Permanence of Paper for Printed Library Materials, Z39.48-1984. Binding materials have been chosen for durability.

Library of Congress Cataloging-in-Publication Data
Adams, John A. (John Alfred), 1951–
    Damming the Colorado : the rise of the Lower Colorado River Authority, 1933–1939 / by John A. Adams, Jr. — 1st ed.
    p. cm. — (The Centennial series of the Association of Former Students, Texas A&M University ; no. 35)
    Includes bibliographical references.
    ISBN 0-89096-426-2 (alk. paper)
    1. Lower Colorado River Authority—History. 2. Water resources development—Texas—Colorado River Region—History. 3. Floods—Texas—Colorado River Region—History. I. Title. II. Series.
TC425.C6A54 1990
333.91′415′09764—dc20   90-32795
                                                                 CIP

To Sherry

Contents

List of Illustrations and Tables   *page* ix
Acknowledgments   xi
Introduction   xv

CHAPTER 1. Floods in Texas: Early Background   3
    2. The New Deal in Texas: The Long Arm of the Federal Government   24
    3. Construction and Conflict: Growing Pains of the Lower Colorado River Authority   43
    4. Marshall Ford Dam: The 1938 Flood   66
    5. Colorado Lights: Texas' Little TVA   93
    6. Conclusion   109

Notes   115
Bibliography   142
Index   156

Illustrations

De L'Isle map   *pages 6-7*
Bluffton Ferry crossing   11
Buchanan Dam contractor at work   18
Foundation preparation for Buchanan Dam   19
Workers prepare to pour footings for Buchanan Dam   20
Vaulted arches of Buchanan Dam   21
Upriver view of Buchanan Dam arches   22
Bureau of Reclamation projects   32
Congress Avenue in Austin flooded by Colorado River   44-45
House on spillway of Austin Dam during flood   46
Buttresses 6 through 13 at Hamilton Dam   53
September, 1936, flood in Austin   61
Workers at Inks Dam   63
Construction rail line at Buchanan Dam   64-65
Alvin J. Wirtz and Lyndon B. Johnson   71
Lyndon Johnson and President Roosevelt   73
Workers sink Austin Dam foundation below limestone bedrock   76
Buchanan Dam, with cableway tower   78-79
Texas delegation lobbying for dam completion funds   82
Cross section of Marshall Ford Dam   84
Marshall Ford Dam construction   89
Lyndon Johnson and Pedernales Electric Cooperative Board   102
Major dams of the Colorado River above Austin   105
Hydroelectric dam schematic   106
Ten-county statutory area of the LCRA   110
Lyndon Johnson   113

TABLES

1. Marshall Ford Dam estimates, 1937   74
2. Lower Colorado River Authority facts, 1941   107

Acknowledgments

As with all such endeavors, this study would not have been possible without assistance and encouragement from many sources.
Foremost I want to thank Larry Hill for his genuine interest in directing my research. He provided keen insights into many aspects of the New Deal and its relationship to events in Texas during the 1930s. I also extend special appreciation to Claude Hall, Henry C. Dethloff, Gary Anderson, and Al H. Ringleb for their counsel and advice.
I gained new respect for the volume as well as importance of primary resource material. Dealing with four collections of presidential papers, the Library of Congress, and numerous national and state archives proved both invaluable and rewarding. Linda Hanson, at the Lyndon B. Johnson Library in Austin, where the Lower Colorado River Papers are housed, gave freely of her time and knowledge. I would also like to thank the Lyndon Baines Johnson Foundation for their generous support of my research. At the Eugene C. Barker Texas History Center in Austin, Ralph Elder was helpful, as were the staff at the Texas State Archives and Biruta Kearl at the Austin History Center. Ken Carter at the Regional National Archives in Fort Worth allowed me to handpick from the vast warehouse of documents to ensure that I would locate necessary data on the Army Corps of Engineers and the Federal Power Commission. Emmett Gloyna, the Texas representative of the Bureau of Reclamation, provided useful background data on the relationship between his agency and the LCRA.
In the winter of 1986, I spent a cold, windswept week at the Herbert Hoover Presidential Library in West Branch, Iowa, assisted by Robert Wood, Dale Mayer, and Shirley Sondergard. Of particular help were Paul McLaughlin, Susan Bozanko, and John Ferris, who each assisted with my research into the heart of the New Deal documents and collections at the Franklin D. Roosevelt Library in Hyde Park,

New York. No study on any key aspect of the New Deal would be complete without a review of the documents at Hyde Park.

Work at the presidential libraries was augmented by numerous trips to the National Archives and Library of Congress in Washington, D.C. Richard Crawford was most helpful in making some semblance of order out of the massive Public Works Administration, Bureau of Reclamation, and Works Progress Administration records in the National Archives.

I would also like to thank Richard Lowitt, whom I, by chance or fate, met in the Manuscript Division of the Library of Congress during the summer of 1986. A recognized expert on the New Deal in the West, he directed particular attention to key items in the George Norris, Harold Ickes, and Woodrow Wilson collections. Our week-long discussions provided valuable insight into the vastness of the New Deal programs in the West and the zealous fervor with which Sen. George Norris fought throughout his lifetime.

My extensive research efforts and travel in addition to managing a multinational company were facilitated by the constant cooperation and encouragement of my family. I would also like to thank Karen Sonley, Loren Shellabarger, Janie Leighman, Ted Hajovsky, Tom Autrey, Jerry Smith, Carl Walker, and J. Malon Southerland, all of whom assisted at various stages of this project. Additionally, I greatly appreciate the encouragement and support of T. L. and June Calvin.

The staff at the LCRA in Austin were extremely helpful. John Williams, public relations director, also knowledgeable in the background of the authority, was a good sounding board for many of my questions and requests. Most impressive, however, were my tours of the dams and power facilities that make the LCRA flood control and public hydroelectric power possible. Joe Irvin and Chris Clymer in the graphics department were a tremendous help with illustrations and photographs. Bobby Bostic, manager of the Buchanan and Roy Inks dams, gave freely of his time and experience of nearly three decades with the LCRA to answer my questions. In like manner, Jim Clayton, also a veteran of over three decades with the LCRA and manager of the Tom Miller Dam, gave me a Cook's tour of the Austin Dam. Going deep inside the bowels of this dam was an experience I will not soon forget. At the Tennessee Valley Authority, John Moulton, in Knoxville, was helpful in supplying information and current data.

I was inspired to dig deeper and longer into the project by a number of people, both past and present. Each in a special way has had an impact on the curious parallels between Presidents Lyndon B. Johnson, Franklin D. Roosevelt, and Herbert Hoover with regard to rivers, rural development, and floods. Although at different ends of the political spectrum from Johnson and Roosevelt, Hoover was one of the foremost authorities in the 1920s on water development, "super power" (electricity) distribution, disaster relief, and the ravages of floods. Hoover's on-site accounts and direction of the flood relief on the Mississippi River in 1927 are most memorable. Johnson, who grew up on the banks of the Pedernales River in Central Texas, and Roosevelt, who was born and raised on the high bluffs overlooking the Hudson River, both had an early and distinctive attachment to rivers. Standing on the back porch of the Roosevelt family home at Hyde Park, with its view of the Hudson valley, I acquired a priceless insight into Roosevelt's early attachment to nature and conservation. With Roosevelt his mentor, Johnson, as a congressman, transformed his own early Central Texas background into a formidable advocacy for flood protection, rural development, and cheap electrical power.

I must note the early contribution of my parents. My mother gave me the drive and cussed hardheadedness to persist against all odds, a trait for which she, although stricken with polio in the prime of her life, is a living example. My dad gave me my earliest remembered interest in history and its impact upon the world around us. Among his many keen historical perspectives were his comments on the impact and importance of the Tennessee Valley Authority on the southeastern region as well as on the nation. Having grown up in the depression era of the 1930s in North Georgia, he vividly stressed in later years to my brother Ron and me how "Mr. Roosevelt's project" had given hope and opportunity to a whole generation of depression-weary people. I will always treasure his counsel as well as his love of history and books.

For both John A. Adams III and Calvin John Green, I appreciate their understanding and patience over the last couple of years.

Finally, and most important, I thank my wife, Sherry, who has been a never-ending source of encouragement and understanding. She has given long hours to helping me prepare this document as well as giving me the flexibility to pursue each and every research avenue. Of all those who have helped, it is she alone who gets sole credit for pushing me over the hump to finish. For her dedication and love I will always be grateful.

Introduction

During the New Deal a number of large multipurpose river improvement projects were either begun or given new life. One such project was the Lower Colorado River Authority (LCRA) of Texas, the culmination of numerous attempts between 1900 and 1933 to harness and develop the Colorado River. The rise of the Lower Colorado River Authority both benefited from other regional projects and contributed to the development of others. Clearly, it profited from the precedent-setting evolution of the Tennessee Valley Authority. Like other major large-scale construction projects of the 1930s, the LCRA was a product of conflict as well as cooperation between state and federal government. This interaction helped settle such issues as the limits of federal authority over navigable rivers, the extent to which states' rights could limit federal jurisdiction, and the resolution of public interest versus private hydropower conflicts. Finally, the development of the Colorado River, by the late 1930s, is an excellent example of one of the many reclamation, conservation, and hydroelectric projects that hallmarked the New Deal in the West.

The roots of the water-development conflict and the eventual delineation of authority can be traced from the turn of the twentieth century. Water development, in large part, was an offspring of the conservation movement generated by young aggressive professionals in many technical fields. These Progressives, as Samuel Hays pointed out in *Conservation and the Gospel of Efficiency*, encouraged "national planning to promote efficient development and use of all national resources." The conservation movement, which included multipurpose river development, exalted the expert technician as the one best equipped to achieve orderly development of the nation's natural resources. In no other region were water and water development so important as in the American West.

The first two decades of the twentieth century witnessed a tremendous federal activity in the arena of water regulation and legislation. Presidents Theodore Roosevelt, William Howard Taft, and Woodrow Wilson all labored to create substantial protection for public domain. The zealous conservationists of the early Roosevelt years were mollified as more and more interest groups became involved with the question of water rights. Taft agreed in principle with reform and maintained the status quo. Wilson, though largely preoccupied with the war in the latter half of his administration, presided over the fruition of a fundamental shift of water rights from state and private interests to federal jurisdiction. Attention then shifted to determining the extent of cooperation with private concerns, which created more questions than answers.

Although the West would prove to be an excellent haven for new industrial development and growth, the emerging electric industry was becoming a dominant factor in both politics and industry, much as the railroads had been some forty years earlier. The decade prior to the New Deal produced mixed responses to multipurpose river development and expansion of conservation in general. The argument for multipurpose river development was expressed by Gifford Pinchot in *The Fight for Conservation:* "To develop a river for navigation alone, or power alone, or irrigation alone, is often like using a sheep for mutton, or a steer for beef, and throwing away the leather and the wool."

Thus, the fundamental concept that the rivers and streams were in the public domain greatly influenced the evolution of water-resource enhancement. Single-purpose exploitation by private industry overlooked the multipurpose approach and benefits to valuable national water resources. Revenue from sales of hydroelectricity could finance a wide range of secondary projects, including flood control and irrigation. Furthermore, the Tennessee River, with its Muscle Shoals rapids, was more than just "water" to Sen. George Norris; it embodied the hope for regional development from its source to the mouth. Unlike other national resources projects, dams properly located could produce power to make such growth and development economically feasible.

The water-development issue was one that could not be disregarded. Water development and the question of private versus public power made front-page news. The nation had been jolted by the sensational Mississippi River floods in April and May, 1927, and the New England floods in November. In these two regions there had been over three hundred deaths, $200 million in property damage, nearly one thousand bridges washed out, and one million people displaced from their homes. Also in the late 1920s and early 1930s, there was Norris's Muscle Shoals bill under debate, a consideration of the proposed Saint Lawrence seaway, and legislation on Boulder Dam before Congress. At the same time, the Federal Trade Commission was conducting an investigation of private utilities, stimulating nationwide concern for the future of inland water development. It was in this environment that the Lower Colorado River Authority was created in 1933.

In less than a decade, during the New Deal, Central Texas initiative with federal government assistance was able to control and develop the Colorado River. Centuries of vast seasonal flood damage were eliminated and exchanged for a well-managed system of dams, reservoirs, and hydroelectric power stations. Initial private efforts between 1900 and 1930 to control and develop the Colorado were thwarted by the massive need for capital and equipment. The economic downturn in the late 1920s and early 1930s dampened most private efforts for river development not only in Texas but also throughout the West.

The depression in Texas afforded both adversity and opportunity. Backed by solid political leadership both in Central Texas and in Washington, a cooperative consensus was developed. As a result, the river was controlled and much-needed jobs were created along with the eventual spread of rural electricity and electric cooperatives. The legacy of flood control and water development on the Colorado in Central Texas was the LCRA as the most extensive multidam project west of the Mississippi River. At the time of their completion the Buchanan and Marshall Ford dams rivaled any in the world. In deference to the earlier Tennessee Valley project the LCRA was fondly regarded as "Texas' Little TVA."

DAMMING THE COLORADO

1.

FLOODS IN TEXAS:

EARLY BACKGROUND

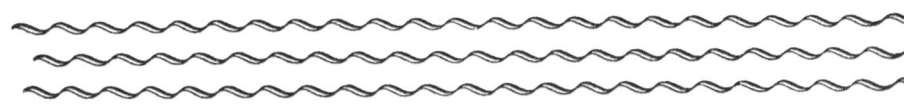

> The 26th, we proceeded on our Journey, and came to the River call'd la Sablonniere, from the many Sand Banks there are in it. The 27th, departing for it, we came to another little narrow river, but very deep.... We ... went to camp beyond it in a little Wood, where we had a very bad night, because of the Rain which fell again, and the overflowing of the River.
> —La Salle, January, 1687

The Colorado River played a distinctive part in the development of Central Texas long before the creation of the Lower Colorado River Authority. The Colorado is an intrastate river system stretching northwest to southeast from its headwaters on the Texas–New Mexico border across the state to its mouth on the Gulf of Mexico. Flowing a distance of over six hundred miles, the Colorado River basin drains forty-one thousand square miles (an area the size of the state of Ohio) and is the second largest intrastate watershed in the United States. With an annual runoff of over two-million acre-feet, the Colorado River watershed is slightly larger than the Tennessee River basin, which flows through seven states, and in Texas is second in size only to the Brazos River.[1]

In addition to the Colorado River, Texas has numerous waterways. There are a dozen major river systems flowing along the borders or through Texas, ten of which are entirely intrastate. Each of these rivers has had its own particular impact on the region in which it is located. The Colorado is no exception. The central location and surrounding countryside have allowed the Colorado a prominence not generally accorded to other Texas rivers. The Colorado passes through three distinct regions of Texas. From its headwaters in Sulphur Springs Draw northwest of Brownfield in the High Plains, the Colorado descends from an elevation of 3,600 feet past the Callahan Divide through Colorado City, Robert Lee, and Ballinger, thirty-one miles northeast of San Angelo. In this area of West Texas there are rolling prairies that drop in elevation from 2,600 to 1,600 feet prior to reaching the Lampasas Cut Plains near San Saba. Running through Travis, Burnet, San Saba, and Llano counties, this central section of the Colorado is the heart of the river. In this region the river is marked by the presence of limestone buttes known as the Balcones Escarpment which intersects with the river at Austin. It is in this upland region, northwest of Austin, referred to as the "Highlands" or Texas "Hill Country," where a majority of downriver flooding originates. Flooding begins in this region in large part because of the narrow canyons, steep slopes, and the rocky nonporous soils that are unable to handle intensive seasonal rainstorms. At Austin the tops of the banks are over 40 feet above the riverbed. With the river dropping from 1,500 to 425 feet in elevation, this region is the location of most of the dam and reservoir development.[2]

The lower third of the Colorado River extends southward from Austin for three hundred miles through Bastrop, La Grange, and Columbus, to Matagorda Bay. This region flows past the Oakville Escarpment into a coastal plain marked by gentle sandy terrain and a meandering riverbed which drops 425 feet to the gulf. Below La Grange the river has a flatter gradient as well as lower banks and a narrower riverbed, all very conducive to flooding. The year-round steady flow of the Colorado below Austin is the result of spring-fed tributaries, primarily the San Saba, Llano, Concho, and Pedernales rivers. Annual rainfall in the area of the watershed ranges from fifteen inches on the High Plains to over forty inches near the Gulf of Mexico. The extensive drainage area of the Colorado basin has largely contributed to the legendary rapid swelling of the river in times of heavy precipitation. On his last exploration to the New World between 1684 and 1687, René-Robert Cavelier, Sieur de La Salle and his party were forced to seek high ground as they encountered the flooding La Sablonniere River, near the mouth of the modern-day Colorado.[3]

Early cartographers of the region depicted the Colorado River by various names and locations. A map by G. De L'Isle entitled "Carte de la Louisiane et du cours du Mississippi," engraved in Paris in 1718 in a format 19 × 25½ inches, labeled the river "Rio de San Marco ou Colorado." De L'Isle, who possibly was the first to use the word *Teijas* (Texas) on a printed map, depicted the Colorado as the largest river in the region between the Rio del Norte (Rio Grande) and the Rio

Rouge (Red River). During the colonial period an English mapmaker, Henry Popple, released from London in 1733 the first large-scale printed map of North America, entitled "A Map of the British Empire in America." There the Colorado, on sheet nine from Popple's map set, is labeled as "R. aux Cannes ou Rio di San Marco o Colorado."

Thus, the well-charted Colorado has served as both friend and foe to explorers and settlers for over three hundred years. The first successful Anglo-American colonizer of Texas, Stephen F. Austin, noted in a promotional pamphlet that "the rivers of Texas are the Sabine, Neches, Trinity, Brazos and Colorado, all navigable a considerable distance into the interior. . . . Of all the rivers in Texas, the Brazos and Colorado are the largest." Twice he mentioned the relationship between surface water and drainage: "The land is sufficiently elevated to drain easily and rapidly after heavy floods or rain." With regard to the nature of the Colorado upriver, Austin assured new settlers that "some part of the northwestern section is hilly, particularly on the head of the Colorado and Guadalupe river. . . . The hills do not form leading ridges that impede the flowing waters to the southeast." No reference is made to flood control, but as early as 1828 Austin did promote water power by noting that "the situation of the streams affords great facilities for water works [grist mills] and irrigation."[4]

Although development of the Colorado River basin progressed at a slower pace than in the more populous East Texas region between the Brazos and Sabine rivers, there were waterpower sites on the Colorado at Goldthwaite, Lometa, Bluffton, Kingsland, and Marble Falls. Dammed for power, the water of the Colorado turned turbine shafts in mills and cotton gins throughout the second half of the nineteenth century.[5]

The state's capital was ultimately located on the banks of the Colorado near the old hamlet of Waterloo and named for Stephen F. Austin. The site was one of two selections made by a commission appointed by the congress of the Republic of Texas. After numerous site inspections, the commission had narrowed the choice to the Brazos River and the Colorado. On April 13, 1839, the president of the Texas republic, Mirabeau B. Lamar, received the commissioner's recommendation:

> We found the Brassos River more central perhaps in reference to actual existing population, and found in it and its tributaries perhaps a greater quantity of fertile lands than are to be found on the Colorado, but on the other hand we were of the opinion that the Colorado was more central in respect to Territory, and this in connection with the great desideratums of health, fine water, stone, stone coal, water power &c, being more abundant and convenient on the Colorado than on the Brassos

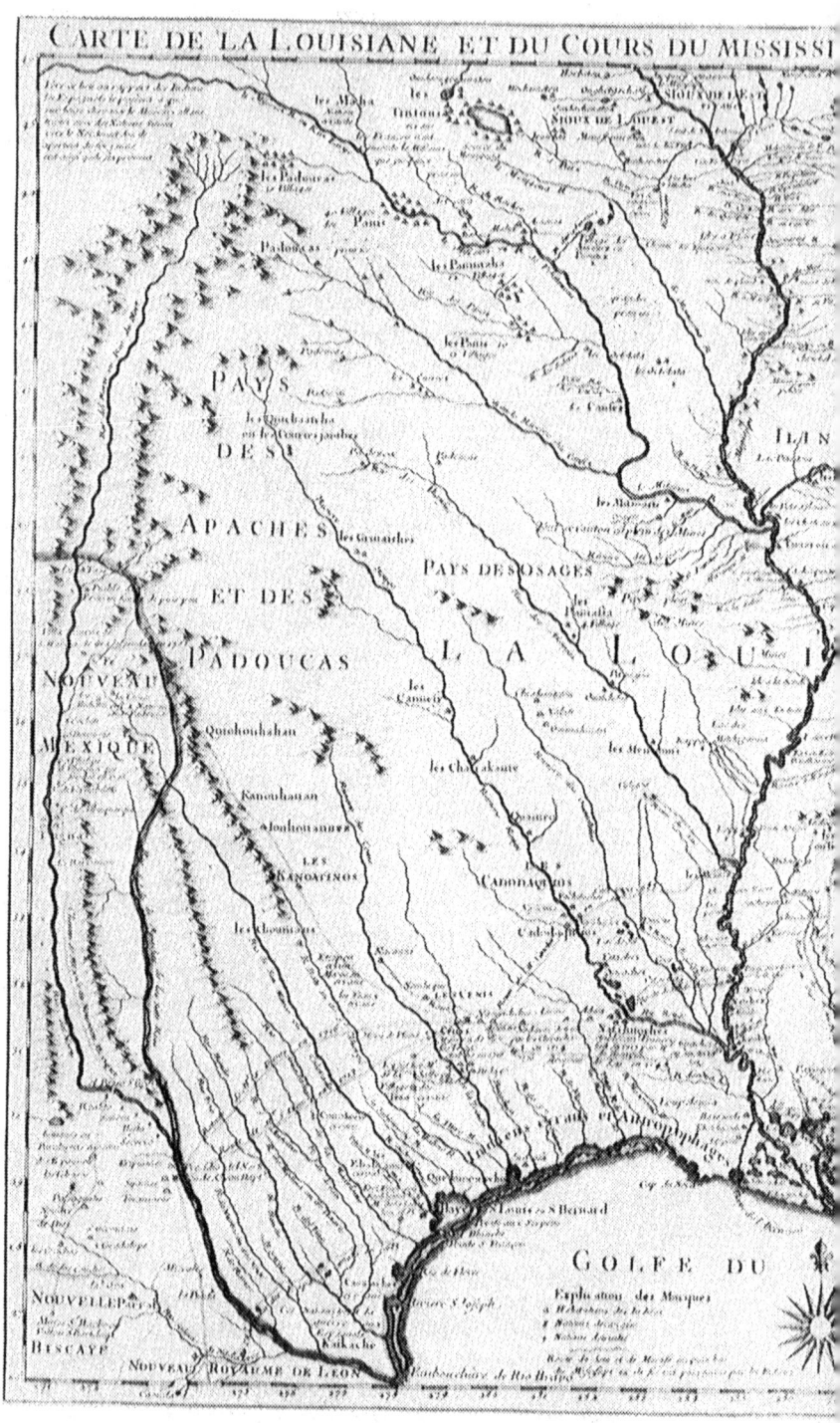

The 1718 De L'Isle map is the earliest rendering of Colorado River.

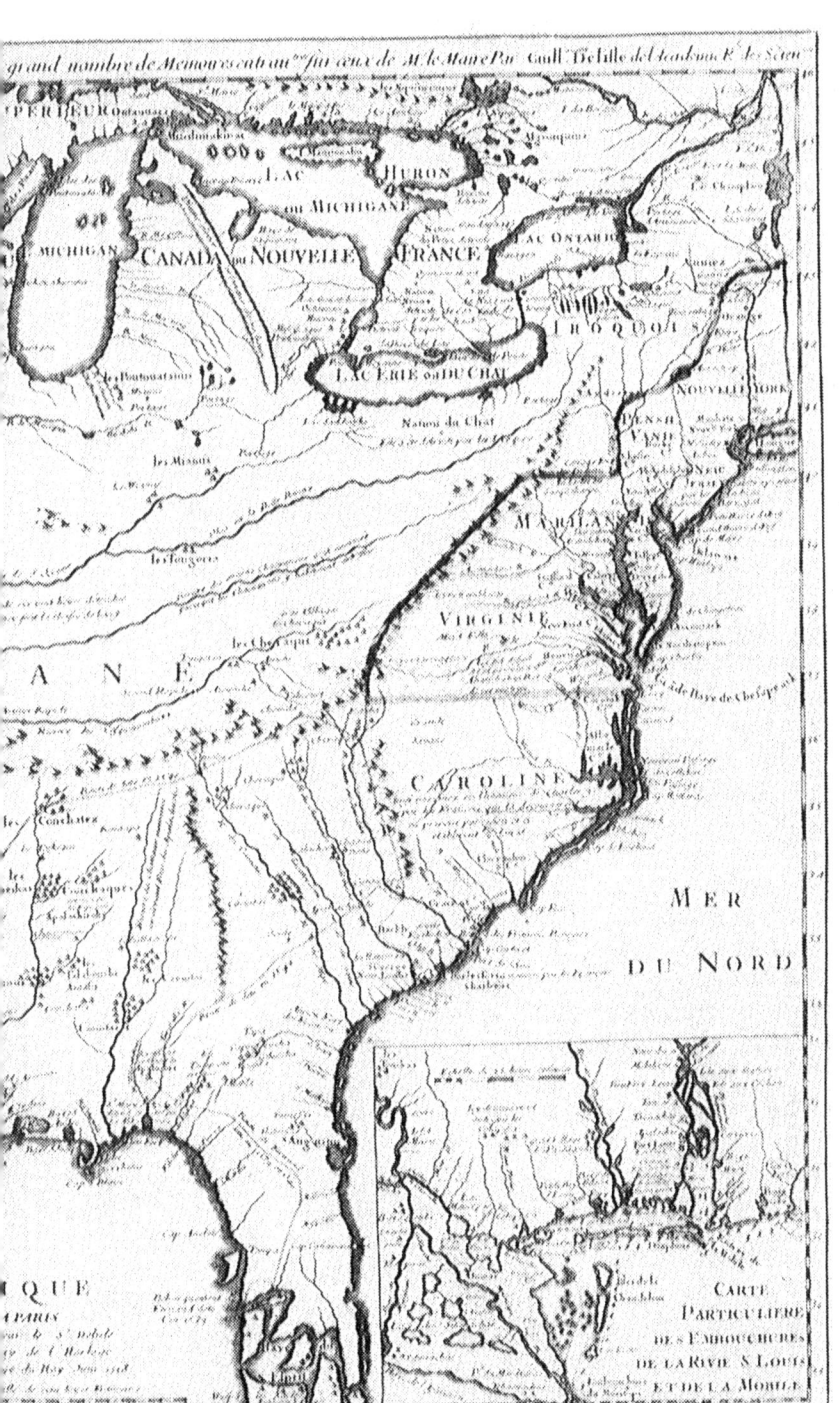

(Courtesy Dorothy Sloan)

River, did more than counterbalance the supposed superiority of the lands as well as the centrality of position in reference to population, possessed by the Brassos River.[6]

The report further stated that the river would provide good transportation, fine water power sites, and a year-round water supply.[7]

Although the report located Austin on the east bank in a central location, it did not contain any reference to flooding, yet such conditions may have been responsible for the river's name. Common belief has established that the river obtained the name "Colorado" or "red" from the early Spanish during a flood or high water. Inundated by red silt around present-day Brownwood, many early Texans referred to the Colorado floodwater as "red rises." Thus, the river quickly gained recognition for its own distinct character.[8]

The frequency of flooding and damage is documented back to the 1830s. Within four years after the selection of Austin as the young republic's capital, the merchants and settlers along the river learned to anticipate annual floods and damage. In February, 1843, the river rose thirty-six feet above normal at Austin, causing tremendous damage and loss of life. Damage downriver from the capital to the delta was even greater. Below Austin the banks were low or nonexistent. Often the river was blocked by natural dams of driftwood and debris known as "rafts." These dams restricted shipping, made water transportation dangerous, and agitated all who lived along the Colorado. By 1870 a combination of city dwellers in Austin and farmers downstream were voicing appeals for control of the river. However, adequate management and flood prevention was long in coming.[9]

The earliest effort to clear and dam the river came from the private sector, beginning in earnest in the mid-nineteenth century. The acknowledged major obstacle to river traffic was the raft of logs and, near the mouth of Matagorda Bay, a natural driftwood dam known as the "Big Raft." In 1851 the Colorado Navigation Company made some headway in cleaning the channel, yet the project was not complete until 1854. Even then the aid of the Army Corps of Engineers was needed. The narrow channel was opened to traders and merchants with the use of $20,000 from the federal government, the first appropriation by Congress for the improvement of the Colorado. Due in large part to neglect, after less than six years, the river was once again blocked by debris. This raft reduced commerce on the river, as did the Civil War and the completion of the railroad to Austin in 1871.[10]

By the late 1880s the rafts were massive. One raft sealed the mouth of the river, the second was inland one mile, and the third was inland about three miles. The length of each was between two hundred and four hundred feet. However, the most massive accumulation of twisted debris was twenty-two miles upstream. It was four miles long, three hundred feet wide, and twenty-five to thirty-five feet deep. This raft protruded thirty feet above the surface of the water at its upper end. Not only was all river commerce to the capital halted but these natural dams also contributed to flooding:

> These rafts act as dams, obstructing the flow of water and, during times of freshet, cause the overflow of many square miles of valuable land. In some cases, the obstructed waters find their way across the country to the valleys of neighboring streams, entailing considerable losses to farmers and others.[11]

Many wanted the benefit of navigation and flood control on the Colorado, but few made any effort to clear the channel. Some local efforts continued into the early 1900s. However, not until after four more major floods between 1900 and 1923 did the state take action. Local reclamation districts in Matagorda and Wharton counties were created in 1924, and had invested over $1.2 million by 1930 to clear the mouth of the river. These efforts proved successful and, with the aid of the Army Corps of Engineers and the Lower Colorado River Authority, the river has since remained open.[12]

Upriver the concern was not with debris so much as it was with continual flooding. Damming the river to regulate its flow became a major preoccupation in the area. As early as 1854 Gen. Adam Rankin Johnson, a local Austin farmer and entrepreneur, had suggested a number of damsites. Years later, in 1885, he attempted to generate support for building a dam north of Austin at Shirley Shoals, the present site of the Buchanan Dam. This site had long been known as a convergent point of runoff that produced flooding downriver. Johnson's recommendation was the result of his entrepreneurial interest in developing the river, which was spurred by the seasonal flooding and his personal inspection of the origins of much of the flooding along the river. However, Johnson's plans never materialized and were not further investigated for a long time. From the 1890s until the early 1920s little was done. Prior to his death, Johnson transferred his land-rights to C. H. Alexander, Sr. In the meantime, flooding in July, 1869, and October, 1870, again focused attention on the lack of control over the Colorado.[13]

The city of Austin began to show more interest in controlling the river as well as securing an adequate water supply and producing hydroelectric power. To accomplish these goals the city began the construction of a dam and generating facilities at the cost of $1.6 million. The generator would have an estimated 1,400 horsepower capacity.

Begun in 1893 and completed in 1895, this bold city project would also supply eight million gallons of water per day to the city.[14]

Initial success of the Austin city dam project quickly eroded. Water running both over and under the dam undercut the "toe" (the leading edge below the waterline) in the downstream limestone foundation. Efforts at repair failed. Work was first hampered in June, 1899, by a minor flood. The plan was to wait until the water level receded enough to place an apron along the leading edge of the dam on the bedrock foundation. Efforts to save the waterpower project and dam were completely dashed on April 7, 1900, when a major flood crested over thirty-three feet, completely destroying the Austin Dam and the power station. One of the most damaging floods on record, the water reached deep into the capital city. Under bold headlines, "The Great Dam Goes Down before the Raging Torrents of the Colorado," the *Austin Semi-Weekly Statesman* reported: "With the breaking of the dam the immense volume of water swept down the river and poured out into the valley below, sweeping houses over like they were tenpins. Within twenty minutes the river immediately opposite the town was sixty feet deep and nearly a mile wide."[15]

In addition to the loss of forty-seven lives, there was a property loss of over $9 million as well as a large indirect economic loss to the region. The superstructure of the Austin Dam, pushed two hundred yards downstream, stood as a glaring reminder of the power of the river and the shortcomings of flood-control planners. From the beginning, the design of the dam and powerhouse were faulty. The unstable limestone composition of the riverbed had not been fully investigated. In an 1898 report by the U.S. Geological Survey it was noted: "The water from the tailrace of the power house flows in front of the toe of the dam and parallel with it," inducing erosion. In a 1904 follow-up study, local Texas water expert, T. U. Taylor, provided the most damaging information: "No hydrographic data had been collected." The reservoir behind the Austin Dam had filled with so much silt that 48 percent of the original lake consisted of mud. In 1893 the storage capacity was 83.5 million cubic yards; in 1900 it was 43.5 million yards. Thus, failure was inevitable.[16]

As the population in the central region of the Colorado increased, so did the flooding. An investigation commissioned by the U.S. Congress in 1930 reported that between 1900 and 1915 there had been a total of twenty floods on the Colorado, four of which resulted in the loss of at least ninety-four lives and caused over $23 million in property damage.[17] Other than helping remove the raft at the mouth of the Colorado to improve navigation, the Army Corps of Engineers had done little to control the flow of rivers in Texas. In nearly a dozen

The growing population of the Austin area made heavy use of the Bluffton Ferry for crossing the Colorado between 1886 and 1911. Fares ranged from twenty-five cents for a rider on horseback to fifty cents for a single team of horses and wagon. (*Courtesy Llano County Museum*)

reports filed between 1891 and 1908, the corps concluded that the development of rivers in Texas was "not feasible," "not worthy," or "not desirable for the United States to undertake improvements."[18] With seasonal flooding continuing to plague Central Texas, however, the mounting waste and destruction—of little concern to the federal government—did not escape the attention of residents along the Colorado.

At the state level both private and public efforts to conquer the river continued during the first two decades of the twentieth century, resulting in a series of developments that set the state for lasting solutions to river-control problems along the Colorado. In 1912 the city of Austin, determined to provide an adequate water supply, undertook plans to rebuild the destroyed Austin Dam. However, these efforts were once again unsuccessful: a second flood destroyed the Austin Dam in April, 1915. Efforts to rebuild the dam had been conducted primarily by private contractors working for the city. These private interests were not able to come to grips with either the technical or financial problems involved in building a safe dam. The city of Austin made demands that could not be met by the small-scale contractors. Though tragic, the second loss of the Austin Dam set in motion greater efforts by both public and private groups to manage and develop the Colorado.[19]

Creation of the Colorado River Improvement Association in August, 1915, marked yet another effort to combat flooding. The most active members were landowners living adjacent to the river in Travis, Bastrop, Fayette, Colorado, Wharton, and Matagorda counties. Their efforts, directed at commercial improvements along the river, included flood control, power development, and irrigation. The rice growers also organized, primarily to ensure ample water during the dry season. The county commissioners along the river "offered to aid the Federal Government in the event that any improvement is undertaken." As a result of the combined efforts and encouragement, the U.S. secretary of war undertook a survey of the Colorado River watershed. The appropriation for the preliminary survey came as a result of the Rivers and Harbor Act of 1916. The primary purpose of the survey was to review, for the first time, previously developed plans by state and county authorities for flood protection and to determine which of these plans the federal government would support on the basis of improving and protecting river "navigation." In the final report to Congress in 1919, the Army Corps of Engineers clearly singled out the intense flood problems of the region. The report recommended a combination of multipurpose dams above Austin and a series of levees below the capital to foster flood control, irrigation, power, and adequate supplies of municipal water.[20]

The numerous setbacks experienced by both the city of Austin and various private interests notwithstanding, the 1919 report concluded that the value resulting from any project to improve "navigation" on the Colorado did not warrant funding from Washington. Although not overly pleased by this report, residents along the Colorado were encouraged to know that the problems of the watershed were at last a matter of record in Washington. The final report stated:

> Conditions at the headwaters of the river suggest the use of reservoirs in conjunction with levees as probably the best means of flood control. There are several local organizations interested in power and irrigation, which with certain State departments, might be expected to *cooperate* [my emphasis] with the Federal Government if it entered upon this work.[21]

The local engineers in the field, though bound by the Corps of Engineers' inclination to improve rivers only for navigation, had at last made a multipurpose survey of the Colorado. By the mid-1920s water transportation to Austin would not have been practical or economical. The rail system and highways linked Austin to all parts of the state. By maintaining a constant hard-line position in favor of navigation improvement, the corps inadvertently fostered more prudent development of not only the Colorado but also all Texas rivers. In the meantime, there emerged a clear recognition of the merits of multipurpose development. Nowhere in the country, except on the Tennessee River, had a project of such magnitude been considered. With the federal government favoring private initiative during the 1920s, the state of Texas waited.[22]

Shortly after the creation of the Colorado River Improvement Association, the state legislature and voters of Texas approved a conservation amendment to the Texas Constitution. Adopted on August 21, 1917, article 16 became the foundation for statewide conservation programs. The new amendment was broad, yet rather detailed in its wording:

> The conservation and development of all the natural resources of this State, including the control, storing, preservation and distribution of its storm and flood waters of its rivers and streams, for irrigation, power and all other useful purposes, the reclamation and irrigation of its arid, semi-arid and other lands needing irrigation, the reclamation and drainage of its overflowed lands, and other lands needing drainage, the conservation and development of its forest, water and hydroelectric power, the navigation of its inland and coastal waters, and the preservation and conservation of all such natural reserves of the State are each and all hereby declared public rights and duties; and the Legislature shall pass all such laws as may be appropriate thereto.[23]

From the mandate of the conservation amendment emerged a wide range of agencies and authorities charged with the protection, control, and development of the state's natural resources. Although the amendment also placed rivers and streams squarely in the public domain, at the time of its passage no objections were raised by the private utilities. Few legislators, or voters for that matter, had knowledge of the full extent of the state's water resources.

In the 1920s there was also an extensive effort on the part of private enterprise to develop a series of multipurpose dams above Austin. Although these private efforts constantly proved to be too little too late, a review of the actions of independent entrepreneurs provides an important background to the New Deal developments and the creation of the Lower Colorado River Authority.

These early efforts to dam the Colorado were initiated by C. H.

Alexander, Sr., who had acquired water rights along the river from Gen. Adam Johnson prior to the latter's death in 1922. For a period of nearly three decades, Alexander held the water rights at key sites along the Colorado. At his own expense he attempted to dam the river in 1909 at Marble Falls, yet was able to construct only a low-level dam about one-third of the way across the river.[24]

Throughout this period no one (public or private) endeavored to coordinate efforts along the Colorado or any other Texas river. The Colorado River Commission, the city of Austin, the Army Corps of Engineers, downriver rice farmers, and upriver city dwellers had their own parochial interests. After the release of the extensive yet unimplemented Corps of Engineers report of 1919, which identified three damsites above Austin, the state made little effort to evaluate the river situation in Texas.

The level of concern over flood control usually grew as a response to citizen indignation over the occurrence of another and still another flood. Extensive inundation across Central Texas in July, 1919, September, 1921, and May, 1922, resulted in statewide meetings to consider plans for flood control.[25] In response to the public outcry, the Texas legislature approved the funding for a statewide study of rivers by the Board of Water Engineers. Working in the shadows of the Army Corps of Engineers, the board (with some cooperation from the U.S. Geological Survey) amassed valuable raw data in 1923 and 1924 that would prove critical in the selection of all future damsites.[26] However, by the mid-1920s, although studies had been conducted, no firm action had been taken and residents along the Colorado continued to wait. Finally, events outside Texas provided the catalyst for developing serious measures to control the Colorado.

During the mid-twenties, problems outside of Texas created additional concern about flood control, a concern that would ultimately benefit the Lone Star State. Destructive floods plagued the Tennessee River basin, the other Colorado River basin (in Colorado and Arizona), and the San Joaquin and Sacramento valleys in California. Experiences in these areas had been very similar to those of the Colorado in Texas, with the sole exception that the Tennessee valley had already begun to reap the limited benefits of multipurpose dams. In 1927 the Tennessee River project, still under pressure from the federal administration to liquidate its holdings, was but a mere infant relative to its future development.[27]

Events transpired in the late 1920s that would change the course of water development in the United States forever. The first and most dramatic occurrence was the massive April, 1927, flood on the Mississippi River that covered over 1,200 miles from north of Saint Louis to the river's mouth below New Orleans. The Army Corps of Engineers was being both shortsighted and naïve when, as early as 1924, 1925, and 1926, it boasted in its annual reports, "the Corps removed all qualifications and informed Congress and the nation that the Mississippi Valley was now made safe from serious flood damage." In 1927 tens of thousands of people were driven from their homes and property damage was in the millions of dollars. Over 18 million acres of the Mississippi watershed was inundated. It was this tragic event that attracted attention to the fact that the Army Corps of Engineers had both underestimated the scope and magnitude of flooding in its concentration on improving navigation. The Mississippi River flood of 1927 caused such tremendous damage to local economies and physical properties along the river that the shock was felt nationwide. A concerned Congress, in a specially called session, reacted by passing legislation in May, 1928, authorizing $325 million to expand the duties of the corps to include flood control.[28]

The primary result of this hasty funding was the extending of levees. Although complete federal participation in flood control did not take hold until the late 1920s and early 1930s, the movement was at least under way. The potential involvement of the corps with flood control was not fully clarified until the General Flood Control Act of 1936. For the first time, an act asserted federal responsibility for controlling floods in river basins nationwide. After 1936, responsibility for flood control was vested in the Army Corps of Engineers.[29]

It is interesting to note that Hoover, who was so against federal involvement with large projects like the Muscle Shoals on the Tennessee River, led the disaster-relief commission to feed, clothe, and resettle those displaced by the flood. Although a consistent hard-liner against federal participation in water projects, he was well versed in the magnitude of effort needed to bring about adequate flood control. Along with others, he slowly began to recognize that large-scale floods were disasters that could not be fully handled by local authorities. While secretary of commerce in 1926, Hoover proclaimed, "water is today our greatest undeveloped resource."[30] The local economies along the river could absorb or protect against the effects of small floods but not large, extensive disasters. This realization by the federal government of its responsibility in providing flood control was to be a key element of New Deal public works activities, which in turn proved advantageous to the development of the Colorado River in Texas.[31]

Another significant event was the beginning of the Boulder (Hoover) Dam Project in December, 1928. No structure or project in the western states had more impact than this gigantic undertaking. Situated on the other Colorado River, between Nevada and Arizona,

Boulder Dam was to be the benchmark for all future multipurpose dams. The overall impact of this project proved a much more valuable example to Central Texas than did the Wilson Dam in Tennessee. Using new technology, the dam was built under budget (a mere $114 million) and ahead of schedule. It fully demonstrated the federal government's ability to construct a major dam—which was officially named for Herbert Hoover in 1947—fifty stories high and signaled yet another step toward closer federal-state cooperation on multipurpose river development.[32]

In the wake of the Mississippi River flood and the beginning of Hoover Dam, the power industry, looking to increase its network of power sources, took a closer look at Central Texas and the rapidly growing market for power. Led by power industrialist Martin J. Insull of Chicago, private companies initiated surveys in mid-1927 with the intent of building five or six dams above Austin at a cost of $15 million. The possibility that hydroelectric dams would also give flood protection captivated interested parties from Austin through the lower valley, since Insull's Middle West Utilities Company was one of the most important energy firms in the country. Yet, in a strange set of circumstances, these high hopes were dashed. West Texas ranchers, led by the West Texas Chamber of Commerce, protested that the Insull project would rob them of their rightful claim to water. A heated debate ensued as upstream interests and downstream advocates, such as the Colorado River Improvement Association and the Garwood Irrigation Company, spoke out in favor of protection against floods and guarantees of adequate water supplies for irrigation during droughts. Caught between these two factions, the Insull company, after spending over $1.5 million for water rights and surveys, withdrew all of its interest from the Colorado in July, 1928. Unbeknownst to the Texans, the Insull empire was financially overextended. Its numerous projects were so interconnected that the economic crash in the late 1920s and early 1930s eventually drove the company into reorganization.[33]

The hasty withdrawal of Insull in 1928 was both a shock and a surprise. However, other interested parties quickly stepped in to take up where Insull stopped. The key figure in the development of the Colorado River for over two decades beginning in 1928 was Alvin J. Wirtz, an expert on Texas water law. Wirtz, an attorney, lobbyist, and former state senator, was the legal counsel for the Garwood Irrigation Company, which supported the Insull project.[34]

Wirtz had watched the activities of Insull with great interest and had become convinced as early as the mid-1920s that, in addition to flood control and irrigation, the Colorado could be a prime source of hydroelectric power. With this in mind, Wirtz encouraged the Emery,

Peck and Rockwood Development Company of Chicago, a power developer and client, to begin at once "to take an interest in the Colorado project."[35] The Emery, Peck and Rockwood company was no stranger to Texas since it was, at this same time, coordinating the construction of a dam on the Guadalupe River. On the advice of Wirtz, in early 1929 the company opened an office in Austin and over the next two years completed the surveys of the river started by Insull and, most important, acquired the necessary water and land rights at six key locations along the Colorado. In order to insure its dominant position, the company exchanged one fourth of the investment stock for the sites owned by the estate of C. H. Alexander, Sr. Wirtz was the key behind-the-scenes political and business advisor for this development. According to Welly K. Hopkins, the man handpicked by Wirtz to succeed him in the Texas senate, "had it not been for [his] original interest, there would have been no Colorado River development."[36]

Acquisition of water rights was not always an easy job. Welly Hopkins, a state senator in the early 1930s, recounted the following tragic incident that demonstrates the tension of the period:

> He [Wirtz] was representing a group of Chicago financial interests in the promotion of the building of certain hydroelectric projects up and down the Guadalupe River, which in fact were forerunners to his interest in the Colorado River development. I had acted at his request as the local counsel in Gonzales in representing him in condemnation proceedings for the lands that were requisite to the basins to be overflowed. From time to time, the representative of the investment house in Chicago, I've forgotten the name of it [Emery, Peck & Rockwood], but it was headed up by a man named Peck, would naturally come down to Texas to view the progress being made. On this given occasion he was in Texas in Wirtz's office. Those kind of projects, while they had their proponents in the community, they had their opponents too—those whose lands were being condemned frequently thought that it was wrong, and the communities got divided.
>
> Well, they were very much divided in Guadalupe County and to some degree in Gonzales County on the issue, and feeling ran very high, particularly when it came to condemnation of one's land. Peck was in Wirtz's office on this occasion, as I learned later. The Senate was in session in Austin; I was called off the floor for an urgent long distance call, and when I took it, the Senator was on the line in Seguin almost in tears. In a rather tragic voice he said, "Welly, they've shot Peck. Come on down here." And I said, "Just as quick as I can get there." I hung up the phone without any further details, got in my little Chevrolet, and drove down there, and they had killed Peck. A man named Holleman whose lands were involved in the condemnation proceeding had come

Private efforts to construct the Buchanan (Hamilton) Dam and hydroelectric generating plant were coordinated by the Fegles Construction Company of Minneapolis. *(Courtesy Lower Colorado River Authority)*

in to see Wirtz at the time that Peck was there, ostensibly telling Wirtz's office help that he wanted to talk about some phase of it. As I learned from Wirtz, they had said, "Well, send Holleman in," and as he came through the door in the inner office where Peck and Wirtz were—Wirtz behind the desk—he pulled out a six shooter and started shooting. He killed Peck almost on sight and tried to kill Wirtz, but he fortunately could get over the desk and wrestle him down and take his pistol away from him.[37]

Once the survey had been completed, and all rights-of-way secured, Emery, Peck and Rockwood in April, 1931, selected the Fegles Construction Company of Minneapolis, Minnesota, to build a hydroelectric generating plant, to be known as the Hamilton Dam, on the original site surveyed by Gen. Adam Johnson some seventy years earlier. Finally, the prospect of building a major dam seemed at hand. The new project, between Llano and Burnet on the Colorado River, would become one of the largest projects in Central Texas, employing over eighteen hundred workers and ensuring over two thousand jobs in the related cement, gravel, and lumber industries. The nearly two-

Foundation preparation at the Buchanan (Hamilton) Dam site in August, 1931. *(Courtesy Lower Colorado River Authority)*

mile-long dam was projected to be 137 feet high and to be completed by mid-1933. But, before new construction started, the Emery, Peck and Rockwood financial firm, in a surprise move in November, 1931, sold all its holdings and the dam project to the Central Texas Hydro-Electric Company, a subsidiary of Middle West Utilities, chaired by Martin J. Insull. The news of Middle West Utilities' renewed interest was greeted with overwhelming approval by politicians and stockholders as well as those along the Colorado. The Insull firm was looked upon as a blue-chip company of the era. Although the depression deepened, it was believed that if any firm could weather the economic downturn, surely Insull could. However, the unexpected did happen. In mid-April, 1932, after investing over $3.5 million building nearly half of the two-mile-long dam, the Insull empire collapsed. Thrown into total chaos, the local workers at the Hamilton site could not believe what had happened. The failure not only caught Texas

Workers prepare to pour the foundation footings of the Buchanan (Hamilton) Dam, October, 1931. *(Courtesy Lower Colorado River Authority)*

by surprise but also sent shock waves through the utilities industry nationwide. Insull stood trial for fraud and embezzlement, and, though acquitted, was deported to Canada. Prospects for the Hamilton (Buchanan) Dam were bleak.[38]

The assets, records, and partially built dam passed into the receivership of Alvin Wirtz. Eager to continue the project, Wirtz worked through the fall of 1932 to secure a new financier. In November, 1932, utilities investor Ralph W. Morrison organized the Colorado River Company. Although financially well off, he could not fund the project without outside support. Aware that the federal government was supporting numerous projects directed at creating jobs, Morrison applied unsuccessfully to the Reconstruction Finance Corporation in early January, 1933, for a $4.5 million loan at 5 percent. With the dam building at a halt, the Colorado River Company in August, 1933, sent a second application to Washington, D.C., this time to the Public Works Administration. Wirtz worked long hours behind the scenes lobbying for a loan to finish the dam. Caught in the federal bureau-

A series of unique vaulted arches were designed for the Buchanan (Hamilton) Dam to ensure both structural strength and continuity of the 1,960-foot-long dam. (*Courtesy Lower Colorado River Authority*)

cracy, the initial Public Works Administration loan was rejected. By late 1933 the prospect for federal help looked dim.[39]

Wirtz, the Colorado River Company, and all Central Texas had hope that the new administration in Washington would look more favorably on funding dam projects. Even though their first two attempts had failed, efforts were still in progress to lift the political "raft" in Washington.

Nearly a century after the location of the capital in Austin, the surrounding region seemed no closer to having a uniform multipurpose flood-control plan for the Colorado. During this period nearly $100 million in property and over a hundred lives had been lost. The stark, uncompleted superstructure of the Hamilton Dam, the failed efforts of private firms to succeed in completing it, and the deepening depression only heightened suspicions that the problem was not soon to be solved. On the eve of the New Deal, the Colorado River maintained its reputation for uncontrolled floods and half-completed dams. Many warned of the potential for future disasters. Settlement

The design and magnitude of the Buchanan (Hamilton) Dam arches are striking in this upriver view in late 1931. *(Courtesy Lower Colorado River Authority)*

along the river had increased nearly 20 percent between 1920 and 1933; furthermore, the state's population of six million in 1933 was slowly shifting from a rural to an urban setting. In 1920 70 percent of the population had been considered rural as compared to under 60 percent by the mid-1930s. The center of commerce and state activity along the Colorado that Texas' President Lamar had envisioned was slowly becoming a reality.

Private companies had both tried and failed, over and over. The mood of the nation was for change. The Mississippi River flood of 1927 dramatized the problem on a national level, and the beginning of construction on Hoover Dam demonstrated that a large river could be controlled and managed. In the case of Texas, these examples as well as the steady demand for flood control and hydroelectric power did not go unnoticed. It was hoped that the nationwide activities could serve as a focal point to draw federal funds to Texas.

Although there were numerous key individuals, the early promoter for the eventual emergence of the Lower Colorado River Authority was the small-town lawyer from Seguin, Alvin J. Wirtz. The lobbying and intrigue he fostered and encouraged in order to control and dam the Colorado pointed toward the realization that federal-state cooperation in the mid-1930s was to be more of a reality than a rarity.

2.

THE NEW DEAL IN TEXAS: THE LONG ARM OF THE FEDERAL GOVERNMENT

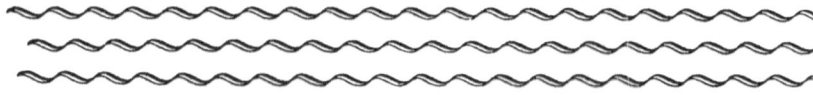

> "Mr. President, I want a birthday present."
> "What do you want, Buck?" Roosevelt asked.
> "My dam," Buchanan replied. "Well, then," responded the President, "I guess we'd better give it to you!"
> —Washington, D.C., June, 1934

One key necessity for the implementation of federal programs on the state level was gaining local political support. In the case of Central Texas, such support was primarily vested in Alvin J. Wirtz. For two decades, beginning in the late 1920s, Wirtz was a local and state advocate and, eventually, a national leader of water development. Born near the Colorado River in the small town of Columbus, Texas, on May 24, 1888, he was preoccupied with developments in the lower river valley for much of his life. After completing law school at the University of Texas in 1910, he practiced with a small firm in Eagle Lake until 1917, when he moved to Seguin on the Guadalupe River. Wirtz, a lifelong Democrat, was elected to the state senate in 1923 and served until 1931. Both as a practicing attorney and senator, he advocated many steps to enhance water development in the state. "Senator" Wirtz, as he was called long after he had left state office, was a specialist in water law.[1]

Wirtz gained his initial expertise in water-rights litigation as attorney for the Guadalupe-Blanco River Authority, which he helped create. In numerous disputes with the city of San Antonio, Wirtz ensured that the small Seguin-based authority prevailed. This knowledge gained during the late 1920s was invaluable for the development of the super project he envisioned for the central Colorado River above Austin. Wirtz proved to be a forceful and talented advocate of water development in Texas. In addition to being an official delegate to the 1928 and 1932 Democratic National Conventions, he was special counsel to the Brazos River Conservation and Reclamation District, and during the early 1940s he served as undersecretary of the interior. In this capacity he gave strong administrative support to the building of the Shasta and Grand Coulee dams.[2]

In the case of Central Texas, Wirtz's primary value to the Lower Colorado River Authority was his ability to operate successfully in the intrigue-filled political environment of Texas during the early 1930s. Because of his acumen he acquired great stature. He was one of a very few in Texas who understood, at this early date, that a change in the presidency meant a "new deal" not only for the country but for Texas.

While Wirtz coordinated the development of the Colorado River in Texas, his close friend, U.S. Representative James P. ("Buck") Buchanan, kept a sharp eye on funding from Washington. Stretching from Burnet in Burnet County to Bastrop and Smithville in Bastrop County, Buchanan's congressional district was split down the middle by the Colorado River. A long-time advocate for control of the river, Buchanan was elected to Congress in 1913. After nearly two decades in Washington, he became chairman of the powerful House Committee on Appropriations. From this pivotal position he exercised political influence in upholding the state's interests in the nation's capital.[3]

Both Wirtz and Buchanan knew from the mid-1920s on that surface-water development in Texas had a direct relationship to other federal reclamation programs in the West. The direction such cooperation would take was yet unknown and untested, but it was clear that the federal government did have an ongoing interest in western water.

Prior to the New Deal era, nearly all water development in the West had been promoted primarily by private sector entrepreneurs. Furthermore, the Newlands Reclamation Act of 1902, which was intended to provide federal funds to augment western water projects, proved effective only in small-scale irrigation projects. With the election of Franklin D. Roosevelt, this trend changed; the federal government took not only an active role but also the lead. This reversal by the federal government resulted from its taking account of two important factors. First, the growing depression had undermined the economic resources of the West; and second, state and local efforts alone could not resolve the major western problems of flooding or developing large-scale projects such as hydroelectric dams. Private operators, either on their own or under municipal contracts, had not been able to grasp fully the magnitude of the tasks involved in planning and constructing multipurpose water projects in the West. So it was that in the 1930s westerners accepted federal aid on an unprecedented scale. Federal funding for key river-development projects provided employment as well as regional development. According to one author, "Government assistance subsidized 'free enterprise' in the West on a grand scale."[4]

The Bureau of Reclamation had carried out the federal government's earlier water-development efforts in the West. Dating back to its establishment as the Reclamation Service in June, 1902, the bureau had concerned itself primarily with reclaiming arid western lands by the development of irrigation projects. In the 1920s reclamation projects involved simple irrigation systems consisting of diversion dams and canals. These early structures usually involved no special large-scale engineering requirements other than those peculiar to each site. In like fashion, early storage projects and dams were generally earthen constructions requiring basic engineering procedures and techniques. Few of these projects were designed to produce power. To augment federal funding, a "reclamation fund" had been set up to accumulate fees from the sale of public lands and small-scale revenues from irrigation leases. To further assist the bureau, Congress had authorized the sale of any surplus power generated by bureau projects.[5] Parallel to its irrigation and storage efforts, the bureau, in the 1930s, had adopted a dual approach for multipurpose dams that would provide both flood control and hydroelectric power. Although the scale of the bureau's projects had changed little between 1902 and the enactment of the Boulder Canyon Project Act in late 1928, by 1930 the bureau had developed the multipurpose concept as a means not only of financing downriver irrigation but also of providing for silt control, navigation improvement, municipal water supply, recreation, fish and wildlife preservation, and salinity repulsion. Even though underbudgeted prior to 1933, the bureau had firmly established itself as the primary water-resource developer in seventeen western states. Thus, Texas conservationists, politicians, and advocates for hydropower had hoped to reap the benefits of its programs.[6]

However, the early promotion of irrigation and conservation in the northern Great Plains and the Far West was given more direct federal attention than projects in Texas.[7] Private and public interests were often in competition. Local irrigation developers were constantly at odds over water rights and jurisdiction. Although Elwood Mead, Commissioner of the Bureau of Reclamation, dreamed of multipurpose projects in the mid-1920s, none materialized until the construction of Hoover Dam.[8] A multipurpose project from its inception,

Hoover Dam across Black Canyon was heralded as much for being a source of irrigation for over 1.5 million downstream acres as it was for being the primary water supply for the city of Los Angeles. Hoover Dam was the first large-scale, multipurpose federal project in the West coordinated by the Bureau of Reclamation. Extending behind the dam for 118 miles, Lake Mead provided a graphic example of the potential for water usage throughout the West. The tremendous size and scope of the project captured the imagination of all Americans: "As the nation languished in the depression, as plant after plant remained idle and company after company went bankrupt, Hoover Dam was being built at a breathtaking pace. The eyes of the country were fixed on it in awe. A landmark event . . . was front page news."[9]

The promise of Hoover Dam, however, could not ease the nationwide impact of the depression. As with the rest of the West, Central Texas sought answers to its economic woes, including unemployment. Public officials did not seem to recognize the full potential of local projects that might put Texans back to work and, thus, prime the economy. As noted, private efforts to build flood-control and hydroelectric projects had failed. The response from Texas administrators in the early 1930s was to support a program of "relief," not reemployment.

Although the Hoover administration had expressed both concern for and encouragement to the unemployed, no massive program or policy was forthcoming to ease their plight. Hoover felt that the responsibility lay with voluntary cooperative organizations and with state and local authorities. He was convinced that a decentralized solution devoid of the federal dole was the key to solving the crisis. In simple terms the depression was to be "mastered by the people themselves and not by the federal government."[10]

Hoover's distaste for enlarging the federal authority precluded a solution during his term of office. He talked only in terms of "indirect relief programs" and voluntarism.[11] Indirect aid, in the form of loans from the Reconstruction Finance Corporation (RFC), increased capital for the Federal Land Bank, and expanded credit from the Federal Reserve Banks, was not forthcoming until well into the second half of his term. By the early 1930s voluntary assistance could meet only about 30 percent of relief demand. Thus, it was left to the states to grapple with unemployment and relief.[12]

The state-sponsored relief program in Texas dated from November, 1932, when Gov. Ross S. Sterling allocated to various counties funds received from the Reconstruction Finance Corporation. This initial system, which distributed relief funds through local chambers of commerce, was changed in March, 1933, by incoming Texas governor Miriam A. ("Ma") Ferguson, who created a central agency known as the Texas Relief Commission.[13] In addition to being short of funds, the commission quickly became entangled in massive red tape, political tampering, and mismanagement. Federal authorities by way of guidelines and directives tried early on with little success to redirect the scope and direction of the state's relief programs. Thus, the initiation of state and federal cooperation between Texas and Washington began with a growing number of federal programs meant to assist the state in hiring the unemployed and stimulating the economy. However, conflict over the administration of the federal relief funds undercut an early spirit of cooperation.[14]

Under New Deal programs, a major change took place as the federal government took a more direct role—creating a multitude of agencies and directing more aid to the state level. The use of grants was a drastic departure from the policy of the Hoover administration. Hoover's belated recognition that the national economy and work force needed government aid resulted in federal action being taken in early 1932. Taking account of the worsening economy and the 1931 report from the president's Emergency Committee for Unemployment Relief, Hoover, in his annual message to Congress of December 8, 1931, asked for the establishment of an emergency financing agency. The prime goal of such an agency would be to *lend* capital to banks, building and loan associations, insurance companies, agricultural credit corporations, farm mortgage lenders, and other private bodies who would in turn *lend* the money to various industries. With the number of unemployed exceeding five million, Hoover recognized the need for more public works and relief. In February, 1932, the Reconstruction Finance Corporation began operation with $500 million in funds and the authorization to borrow up to $2 billion by means of tax-exempt bonds.

In July, 1932, shortly after Franklin D. Roosevelt was nominated in Chicago to run for president on the pledge of providing "a new deal for the American people," Hoover signed the Relief and Reconstruction Act slated to enlarge the programs of the RFC, primarily by increasing the money it could loan to $3 billion. Furthermore, funds could be lent directly to state and local agencies for supporting relief and public works, provided that those agencies matched the dollar amounts borrowed from the RFC. With too few states having the ability to borrow, Hoover's program provided minimal relief.

On the eve of Roosevelt's election the unemployed exceeded thirteen million in number and business losses were estimated as high as $6 billion. Banks were closing and agricultural prices were in a steady decline. After his inauguration Roosevelt took immediate action. With unemployment totaling over one-quarter of the nation's work force by early 1933 and growing (fourteen million people without jobs), he encouraged Congress to pass the Federal Emergency Relief Act. This new legislation in 1933, unlike the Hoover approach, allowed immediate *grants* for state relief. More important, millions were to be set aside for self-supporting projects.

Thus, with the infusion of money from the Federal Emergency Relief Administration (FERA) via the Public Works Administration (PWA), Texas was encouraged to coordinate projects and activities that created jobs. During late 1933 Lawrence Westbrook directed the Texas Relief Commission in Austin. Westbrook provided both timely observations on the Texas economy and maintained a close working relationship with Harry L. Hopkins and Aubrey W. Williams, the primary directors of FERA in Washington. This exposure would prove vital to future river development in Texas. Westbrook's candid reports ensured that conditions in Texas were represented in Washington: "Unquestionably economic condition in these areas [of Texas] have been very bad, but there has been considerable improvement from rain which fell in time for feed crops."[15] Although realizing the magnitude of the depression in Texas, Westbrook always seemed to project optimism for an eventual recovery.

Recognizing the political and administrative problems, Westbrook worked to shift the over $2 million received from FERA away from direct *unemployment relief* to work *employment projects* coordinated by the Civil Works Administration. The CWA was established by executive order in November, 1933, with Harry Hopkins as its administrator. Its immediate goal was to provide approximately four million jobs over the winter of 1933–34. As a transition employment agency, the CWA ceased operations in March, 1934. There seemed to be no limit to the variety of projects supported during late 1933 and early 1934. Employment in any endeavor was the goal, thus justifying use of relief funds for such Texas projects as tick eradication, beautification of the Alamo, sanitation improvements to control typhus and malaria, construction of a new Sabine Pass Post Office, and the elimination of citrus canker in Galveston County (even though there were only nineteen citrus trees in the entire county). The administration of such funds was a very sensitive political task since each county and city demanded its fair share. Thus, while cooperating with federal administrators to fund numerous projects, Westbrook became entangled in Texas politics. Lessons learned during this period proved beneficial to those who later set out to raise funds for the Lower Colorado River Authority.[16]

Several factors combined to produce strained relations between

Westbrook and state politicians. Many state leaders distrusted his outspoken cooperation with federal officials. Gov. Miriam Ferguson wanted more control over the Texas Relief Commission and, therefore, was constantly at odds with Westbrook. In order to serve their own political ends, local politicians wanted not only more money but also more control over the FERA allotments. Yet, the most heated opposition was generated over Westbrook's plans for the unemployed, primarily in urban areas. Influenced by Aubrey Williams of the FERA, Westbrook drew up plans for a subsistence colony in Texas, which he termed "a solution for the technologically unemployable who are on the relief . . . rolls."[17] Westbrook's plan called for moving the unemployed out of the cities into "quasi-municipal corporations" or colonies. The Public Works Administration (PWA) would provide funds to build new dwellings and make improvements and the Agricultural Credit Corporation would furnish credit for livestock and agricultural assistance. Westbrook advised Hopkins that "selection of colonists could be made from civil works rolls [of those without jobs], and thus the CWA program gradually brought to a logical and satisfactory conclusion."[18] Westbrook's initial colony plan for Texas was denied in late 1933, but colonies were established in Texas in the mid-1930s after Westbrook's departure from the state.

Westbrook's control of the Texas Relief Commission and its funds ended in late December 1933. In a confidential letter to Aubrey Williams, Westbrook, whom some had been considering a possible future candidate for governor, stated that "without question there is a serious and important conspiracy underway to get control of Relief and Civil Works Administrations machinery in this State for political purposes."[19] His concerns were well founded.

Just prior to New Year's Eve, Westbrook again advised Williams in Washington that he had discovered plans of "the conspirators to oust me . . . to the point that United Press [news service] had already been given the story." Federal assistance could not and did not help what Westbrook called a conspiracy: "Politicians of the state and the interests which dominate them have evidently come to the conclusion that control of this organization is the most important political factor in the entire state, and will stop at nothing to gain control."[20] Westbrook further said that both his home and office telephone lines were "tapped."

Thoroughly frustrated, Westbrook resigned from the Texas Relief Commission in late January, 1934, and Texas relief efforts remained unorganized as the depression deepened. The number of unemployed cases on the relief rolls totaled 147,174 statewide. During 1934, confusion continued in Texas as state politicians could not come to terms with federal efforts.[21] Detailed FERA reports from Hopkins to all state emergency relief administrations outlined "suggested projects" for work relief. He suggested fields of activity within which all work projects should fall: "planning of public works, housing improvements, production and distribution of goods needed by the unemployed, public welfare and public education."[22]

During the early 1930s there was little cooperation between state and federal officials. Federal agencies were constantly at odds over the handling of federally sponsored programs in Texas. For example, Hopkins strongly encouraged state officials in 1934 to select workers on the basis of need: "No person shall be employed less than *fifty-four* hours a month nor less than *three* days in any one week."[23] Counter to federal guidelines, Texas relief administrators interpreted their mission from a different perspective. In late March, 1934, a letter directed statewide to county administrators, the Texas Relief Commission, now directed by Adam R. Johnson, Jr., concluded with this attempt at clarification: "There seems to be a mistaken idea as to the nature of the program which is before us. . . . Keep in mind definitely that the program is ONE OF RELIEF AND NOT OF EMPLOYMENT. . . . Relief is what we are primarily interested in."[24]

Clearly, the federal-state relationship would have to be more compatible and the conflict over project selection, work hours, and employment rolls would have to be ended before large amounts of funding would be allocated to Texas. Advocates of the Lower Colorado River Authority realized that such friction would undermine the cooperation needed to develop the Colorado River.

Political maneuvering in Texas, as can be seen in the case of Westbrook and the Texas Relief Commission, often moved in crosscurrents with federal intentions. Response to the New Deal in Texas, as in the remainder of the West, concentrated on creating jobs to provide some degree of stimulation to the local economy. However, once the intent of the federal government was either understood or accepted, it was left to Texas to comply and cooperate with Washington in order to receive large-project funding.[25]

During the uneasy transition of the federal government from Hoover to Roosevelt between late 1932 and early 1933, the only water project west of the Mississippi that proceeded unchecked was the construction of Hoover Dam. Prior to 1933, funds were doled out for reclamation projects based primarily on receipts coming to the reclamation fund from public-land sales, oil royalties, and repayment by water users. During a period of thirty-one years the Reclamation Service had expended approximately $227 million to finance thirty-one irrigation projects embracing over two million acres. In 1933 alone

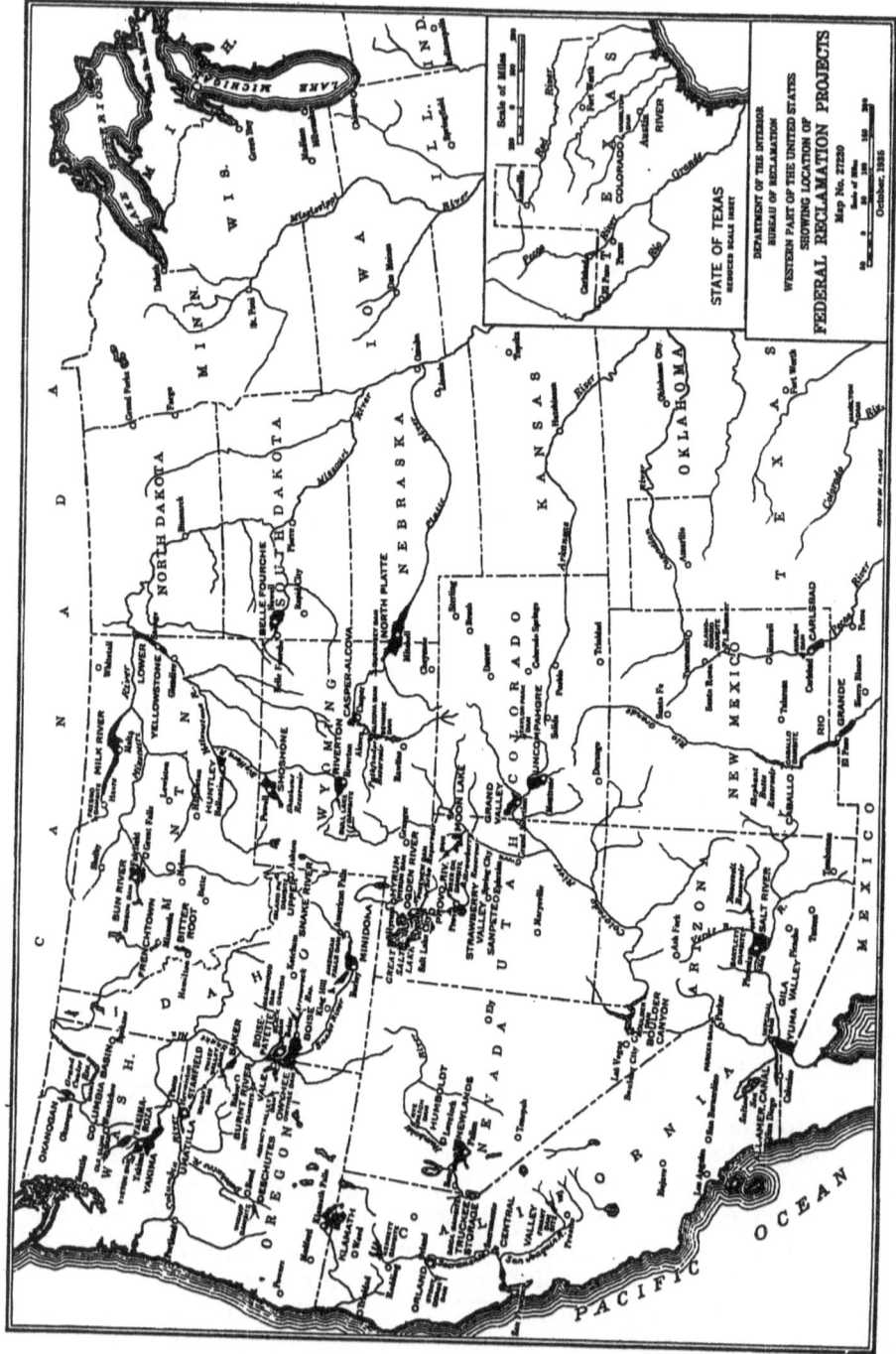

The major projects of the Bureau of Reclamation in December, 1935. The Buchanan site is labeled as Hamilton Dam. (*Courtesy Bureau of Reclamation*)

the total funding provided for reclamation projects by the Public Works Administration equaled the total of the three previous decades. The federal government had firmly assumed the responsibility of substantially financing all future western water projects.[26]

The thrust of these New Deal efforts came directly from Washington. An extensive report prepared for Secretary of Interior Harold Ickes by two nongovernment consultants concluded in late 1934 that two fundamental changes were needed in water-resource development. First, the reclamation projects spawned by the Newlands Reclamation Act were inadequate. Small-scale, inexpensive projects could not be depended upon to serve the needs of the West. Thus, only large, expensive projects should be planned for the future. Second, developing new irrigation projects from existing streams without dams would be difficult because the West was quickly approaching the exhaustion point of surplus surface water.[27]

Further development of large irrigation projects evolved into more multipurpose programs that included flood control and production of hydroelectric power. Although Ickes was at first hostile toward reclamation efforts, he slowly changed his views after realizing the tremendous potential for generating hydroelectric power. He surmised that water storage, when used to generate power, would have more long-range benefits than irrigation would. Thus, hydroelectric projects in conjunction with full river-basin planning was by the late 1930s openly endorsed by Ickes. Hydroelectric power development and the allocation of funds to the various western states enabled Ickes, who also was head of the Public Works Administration, to accumulate great political power within the Department of Interior and the PWA during the New Deal. And Texas stood out as a prime western candidate for federal assistance to fund river improvements.[28]

The sheer magnitude of available federal funds caused political leaders in Texas to reassess their past dealings with Washington. The need for relief funds dictated that they avoid the conflicts of the past. A foremost concern in the minds of local Texas politicians was what they would have to give up to secure such funding. Beginning in 1934, a closer working relationship developed between numerous state officials and Texas politicians serving in Washington. Encouraged by former Texas senator Alvin Wirtz, Representative James P. Buchanan, C. G. Malott (president of the Colorado River Company), John A. Norris (who was chairman of the State Board of Water Engineers), and representatives of the Colorado River Company met with Secretary of Interior Harold Ickes and his staff. In allotting relief dollars Ickes was second in importance only to FERA director Harold Hopkins. To spearhead the evaluation of funding for the lower Colorado River project in Texas, Ickes depended heavily upon the general legal counsel of the Interior Department, Henry T. Hunt, and L. H. Mitchell, a Denver-based reclamation engineer. Some have pictured Ickes as being unsympathetic toward reclamation projects, yet his support of the development of the lower Colorado River argues to the contrary.[29] In a cryptic message to Elwood Mead, commissioner of reclamation, on January 9, 1935, Ickes requested "a thoroughly competent engineer who can make an investigation of a proposed PWA project in Texas."[30] Ickes thus set into motion what was to be the first major federal cooperation with the state of Texas in response to the longtime efforts to control and develop the Colorado River for beneficial purposes.

It was Ickes's desire, after numerous requests from Representative Buchanan and private interests in Texas, to determine the economic and engineering feasibilities of a major project on the Colorado River in Central Texas. Without knowing which area of Texas Ickes had in mind, Mead, who had assumed the investigation would concern the Rio Grande Valley, recommended two engineers, W. E. Anderson of San Benito and L. H. Mitchell, a longtime Reclamation Bureau man in the Denver office. Ickes, in turn, personally picked and briefed Mitchell to function primarily as an investigative engineer for the purpose of determining the potential of the Colorado.[31]

As the engineering evaluation proceeded in Texas, Buchanan made sure funding was available. Even though early requests for funds to help complete Hamilton Dam were denied by Secretary Ickes, Buchanan, who represented the Austin district, appealed directly to the president. Roosevelt was sympathetic, yet outlined one condition. Texas had the option to establish the proper public agency (authority) or forgo federal support. As simple as the choice might seem, the Texas response was cumbersome, resulting in delays by local state politics.

The creation of a public agency to handle the funding of the lower Colorado River did not at first seem complicated. The person in Texas most concerned, Alvin Wirtz, worked rapidly, requesting Buchanan to make an effort to cut the Washington red tape. There were a number of precedents to support creation of a new authority along the midreaches of the Colorado. Under article 16, section 59, the state constitution allowed for an agency such as the president requested. Indeed, four such public bodies had earlier been established in Texas. In 1929 the Brazos River Conservation and Reclamation District had been founded, followed by the creation in late 1933 of the Guadalupe-Blanco River Authority, the Valley Conservation and Reclamation District, and the Lower Neches Valley Authority. Development of an agency for the lower Colorado River did not proceed smoothly, however. A number of factors, the least of which was that the Colorado flowed through the heart of the capital, had much to do with both the delays and the eventual creation of the Lower Colorado River Authority.[32]

As the court-appointed receiver for the defunct interests of both the Central Texas Hydroelectric Company and the Mississippi Valley Improvement Company (both former Insull companies), Alvin Wirtz had moved quickly to promote establishment of the private transitional firm, the Colorado River Company (CRC). As a result of the CRC's initial funding request being denied by the Public Works Administration, Wirtz (even before Roosevelt made his pledge to Buchanan) helped draft and introduce a bill in the Texas legislature in October, 1933, calling for the creation of a "Colorado River Authority."[33] The bill passed the senate but reached the house too late in the session to be considered. A second attempt, made in early 1934, also failed. Efforts to push a bill through both houses were simply not organized well enough. One source concluded that these first two bills lacked support because many legislators held out hope that the Colorado River Company would receive federal funding and not need a state agency. Apparently too many local politicians were still not ready to accede to the federal guidelines on funding.[34]

Investors in the CRC, as well as many Texas legislators, wanted federal money and control vested in a privately owned company. In June, 1934, a compromise was reached. The PWA granted an allotment of $4 million to complete the Hamilton Dam and reservoir. The allotment, contrary to initial PWA guidelines, was a direct loan to the Colorado River Company. However, the money was not given without restrictions. The carefully worded announcement made by PWA administrator Harold Ickes stipulated that the Colorado River Company of Texas "will be utilized as an agency to carry on the work [on the Hamilton Dam] pending creation by the Texas Legislature of a public authority to take it over."[35] In order to exercise effective federal control, Ickes required that he be allowed to name a majority of the nine-member board of directors of the Colorado River Company. As it turned out, Ickes named five members, the Texas State Board of Water Engineers named three members, and stockholders of the existing company chose only one representative. Before the allotment was transferred to the company, the new board agreed to give first option to the state of Texas, or a public body created by it, to purchase at a fair market rate all the property and rights belonging to the company. With Wirtz operating in Austin and Buchanan lobbying in Washington, the Hamilton Dam project had at last secured funding. Central Texas was excited over the turn of events. It was projected that completion of the dam would take eighteen months, employ 1,348 men, and provide a tremendous boost to both the morale and the economy of the region.[36]

With the first major step under way, Buchanan admonished the state legislature to fulfill its side of the bargain by creating an authority for the development of the Colorado River. In a public statement at the Hamilton Dam in mid-July, 1934, Buchanan assured the large audience that much more funding would be forthcoming from Washington once the public authority was established. It was the goal of both Buchanan and another Texas congressional representative, J. J. Mansfield, who also represented a lower Colorado district and was chairman of the House Rivers and Harbors Committee, to lead the way in promoting at least four dams northwest of Austin. During the mass meeting at the Hamilton Dam, sponsored by both the Austin Chamber of Commerce and the Colorado River Improvement association, it was unanimously agreed by the enthusiastic crowd to rename the dam in honor of Congressman "Buck" Buchanan. The crowd had no authority whatsoever to rename the dam. Federal law dictates that a dam cannot be named for a "living" person. This, technically, was of no concern to those in Central Texas.

Sensing the opportunity ahead for all Central Texas, the excited Buchanan assured the crowd that the Colorado River project would be both "the hydroelectric power center of Texas" and "the biggest thing next to T.V.A."[37] Buchanan's dreams for development of the Colorado, which dated back as far as 1915, now seemed reality.

Meanwhile, Wirtz labored long hours to have the district court terminate the receivership of the former Insull properties and water rights so the Colorado River Company could resume construction. To further complicate matters, the transfer of property rights was questioned for the first time by the private utilities companies serving Central Texas. Faced with competition from a public agency, these privately owned enterprises were prepared to fight completion of the proposed dams on the Colorado River.[38]

By late August, Wirtz was able to secure the release of all properties and rights in order to transfer them to the Colorado River Company and its president, C. G. Malott. With legal matters resolved for at least the short term, Wirtz turned his attention to a third attempt at securing legislation to create a state authority. At a dramatic joint house and senate committee hearing on September 5, 1934, former Senator Wirtz arranged to have favorable testimony presented by a Washington delegation made up of Representatives Buchanan and Mansfield and Henry T. Hunt, chief legal counsel for both the Interior Department and the PWA. The Washington delegation remained in Texas through mid-September and continued to offer testimony and support. After a few minor amendments concerning water rights, the senate passed the Lower Colorado River Authority Bill on September 17 and forwarded it to the house for consideration, where it met opposition from the large private utilities, now joined by a group of West Texas ranchers.[39]

Supported by the private utilities companies and the West Texas ranchers, Representative M. V. Dean of San Saba offered an amendment to the senate bill stipulating that all water rights of the new authority be subordinate to the claims of towns within the watershed of the river. Given broad support, this amendment was approved.[40] Without revealing their opposition to public notice, the private utilities in Texas lobbied to amend the bill. And to the surprise of the utilities, and possibly as a result of their machinations, the most heated opposition to the LCRA Bill was led not by the West Texans but by Representative Sarah T. Hughes of Dallas, who insisted that an amendment be added in order to guarantee that no one individual would gain after the liquidation of the CRC and transfer of assets to the LCRA. Hughes was pointedly referring to Ralph W. Morrison, who was a majority stockholder in the Colorado River Company.[41] Additionally, Morrison owned public utility companies in both Texas and Mexico, was part owner of the contested Insull properties in Texas, and was the second largest contributor ($100,000) to Roosevelt's 1932 campaign. Hughes's amendment to block Morrison passed; however, the House refused to concur with both the Dean and Hughes revisions, thus killing the bill in late September.[42]

Although exhausted, supporters of the LCRA pressed Governor Ferguson to call a special session of the Forty-third Legislature. A strong advocate of the river-improvement concept, the governor did so on October 8, 1934, directing the legislators to consider six items, one of which was a new LCRA bill. By now the lines were firmly drawn. Those forces in favor included a cross-section of lower-valley rice farmers and irrigation companies, investors in the former Insull project who hoped to recoup part of their investments, and citizens who saw the development of the Colorado River as a multipurpose undertaking that would not only provide flood control, power, and irrigation water but also massive funds from Washington to provide jobs and relief for the region. In opposition were the West Texas ranchers and their chambers of commerce who were concerned about future water rights on the upper one-third of the Colorado River watershed. The public utility companies were particularly vocal, as were "a few individuals who objected on moral grounds to the financial manipulations of the Insull Companies and of the CRC and contended the state government *should not* have anything to do with the project."[43]

The most vocal opposition, however, came from the West Texas ranchers. Their water and the rights to it were not up for compromise.⁴⁴ They objected to the PWA's insistence that all water rights above the dams be assigned to the LCRA. Thus, the agitated West Texans served the interest of the public utilities companies quite well. They hoped that the water-rights issue would draw attention away from their own selfish interests. This action was reminiscent of the utilities' opposition to the Muscle Shoals project in Alabama and to Hoover Dam in Nevada. Due to provisions in the state constitution, the utilities had already realized they could not fight creation of the authority on the grounds of private ownership versus state government in the power business. However, the collective opposition did not fully gauge the changing political climate in the state.⁴⁵

Governor Ferguson, a clever politician, realized that passage of the LCRA resolution would bring substantial federal funds to Texas; therefore, she gave this measure top priority in her dealings with the special session of the legislature. A fourth bill, providing for an authority with jurisdiction in ten Central Texas counties, was passed by the senate on October 17 by a vote of twenty-seven to one. Once the bill was in the house, the West Texans renewed their campaign for a just and fair water-rights amendment. The West Texas Chamber of Commerce moved its office to Austin in order to concentrate its lobbying effort on a number of issues. To complicate the proceedings further, West Texas representative Metcalf recommended that an "Upper Colorado River Authority" be established to protect the northwest portion of the watershed. Appeals by key West Texas Chamber of Commerce officials were also made directly to President Roosevelt. Upon receiving a noncommittal response, not from the president but from H. T. Hunt, the legal counsel of the PWA, the West Texas lobby only intensified its efforts.⁴⁶

By late October the House and Senate Conference Committee was again at a standstill over state Senator Dean's water-rights amendment and a modified version of the Hughes amendment. Anticipating another deadlocked bill, Buchanan, himself lobbying in Austin, hastily decided to seek a solution in Washington. After extensive negotiations, the Public Works Administration reversed its firm stand against the West Texas lobby and agreed that the water-rights amendment could be accepted. Other than referring the issue to the courts, nothing could be done concerning Morrison's interests. Buchanan's dramatic telegram of November 8 to Sen. Walter C. Woodward, president pro tempore of the Texas senate, clinched the final passage of the Lower Colorado River Authority Bill two days later:

I conferred at length with Hunt and Burke of PWA on the Colorado River Authority Bill now pending in the Legislature of Texas. Stop. As you know I favor the development of Texas and every section thereof therefore I want the Colorado Authority Bill to deal fairly and justly with the people throughout its watershed. Stop. Believing as I do that abundant floodwaters go down this river which if conserved will meet every demand of municipalities irrigation and production of hydroelectric power and further believing that I can procure the allotment of the necessary funds to complete the Buchanan Dam with Dean or Public Policy Amendment in the bill. . . . If the legislature will pass this bill will ultimately allot ten to fifteen million to complete the entire project.[47]

Buchanan was elated with the turn of events. Yet, his two decades of involvement in trying to develop the Colorado River caused him to urge extreme caution. The final paper work was not yet complete in Washington. After the passage of the Lower Colorado River Authority Act, he contacted Austin representatives and "urged quick movement, dollars to run out under PWA."[48] Given his vantage point in Washington, Buchanan knew from earlier experiences in dealing with Secretary Ickes that the initial victory of establishing the Lower Colorado River Authority would be dashed if there were no remaining PWA funds. In late February, 1935, Buchanan received assurance from Ickes and Hunt that everything possible would be done to assist the Colorado River project. Ickes informed his assistant, Hunt, that "Congressman Buchanan called me on Saturday and I told him I would assign you to the task of clearing the ground of all preliminary work . . . [and] to pave the way to an early undertaking of this project."[49] Ickes's primary concern over LCRA funding was the outstanding liens totaling approximately half a million dollars from the defunct Insull ownership. Not only were the dollars important but Buchanan also fully realized that the new authority would "constitute the first precedent in the development of intrastate rivers for any purpose other than navigation by the Federal Government in the entire Southwest."[50] Unlike such multistate agencies as the TVA and the Colorado Compact, the LCRA constituted singularly an Austin-Washington relationship.

With the passage of the act, there remained an uneasiness over federal government funding for the Colorado project. Since the LCRA had been promoted on the basis of flood control and irrigation, those in favor of it temporarily sidestepped the hydroelectric-power issue. There remained many skeptics in both Washington and Texas. In theory federal money could be expended for flood control, irrigation, channel improvements, and dams, but could federal money be used to finance a public authority that would provide cheap power in direct competition with local private interests? The Muscle Shoals and the TVA experiences did not bode well for the future of power generation by the LCRA. From Washington, Elwood Mead expressed his concern regarding this dilemma:

> The Secretary [Ickes] asked for my opinion of the possibilities of the Colorado River project in Texas.
>
> There appears to be no question about the benefits which might come from conserving the water of this river for power and irrigation, but realizing the importance of the kind of an organization, which is to handle this power development, I'd like to know what it is, if it has been worked out.
>
> I know something of the climatic vicissitudes of that region. I don't think any definite revenue from irrigation can be anticipated unless it is based on an organization of the irrigators, contracting to take a definite quantity of water at a definite price, the contract being secured by a deposit of the stock of the irrigation company, as has been done in the Utah irrigation development; or by a lien on the land, as has been done under most of the arid State irrigation district laws. It sums up to this: That the problem is not of the resources of the land and water, but the financial organization which is set up to develop the property.
>
> I take it that the $4,500,000 to complete the dam does not include money for a power installation. Unless there are some conditions about which I am not informed, it seems that the safer plan would be for the government to install the power plant and operate it, and sell the current at the switchboard.[51]

The power companies in Texas did not abandon the fight even after creation of the LCRA. Texas Power and Light lobbied for a full review of the new bill. Sarah Hughes continued to appeal directly to Ickes, urging him to reject all applications from the authority. The vocal Dallas legislator maintained that it was not a legal body. Two developments aided the opposition. The authority itself could not take action as a legal entity until ninety days after passage of the law. Moreover, the initial grant of $4 million allocated by the president in mid-1933 had been expended. As yet there was no accounting for the expenditures. The sum total of these circumstances concerned Secretary Ickes.[52]

In order to reduce involvement in a state project that was not in the best interests of the federal government, Ickes conducted an extensive investigation during January and February, 1935. He appointed L. H. Mitchell, Henry Hunt, and Elwood Mead, each charged with certain responsibilities, to explore the financial and legal status and engineering feasibility of the half-completed Hamilton (Buchanan) Dam as well as the prospects for other dams and reservoirs. Hunt handled legal matters while Mitchell made all site inspections. For example, Mitchell made automobile trips covering a distance over seventeen hundred miles both above and below Buchanan Dam to interview local residents and survey the river. Mitchell and Hunt each compiled a detailed report. Various data were provided by C. G. Malott of the Colorado River Company, John Norris of the Texas State Board of Water Engineers, and E. B. Debler, hydraulic engineer of the Bureau of Reclamation. Buchanan and Wirtz sought to clarify questions Ickes might have concerning the expenditure of the initial $4.5 million. With the preliminary investigation complete it was estimated in early 1935 that about $20 million in new funds would be needed to complete Buchanan Dam and build a series of dams and irrigation facilities farther downriver. The final assessment by Hunt highlighted flood control and irrigation and made only minimal mention of the prospect of hydroelectric power.[53]

At last, on May 7, 1935, in a confidential memo prepared by Henry Hunt in his capacity as chair of the Advisory Committee on Allotments, a resolution was prepared for the president to approve $20 million for the Lower Colorado River Authority. In this recommendation, P.W.A. Docket no. 380, justification for the funds was based on "completion of [the] incomplete dam, reservoir and other works" at the Buchanan damsite. The wording in the letter to the president was general enough to include approval of a "unified system and series of dams at and below that site, impounding reservoirs, hydroelectric works, works for irrigation and other works" to include power transmission lines. This wording was critical to future development and funding for the LCRA, which stretched well into the early 1940s. In their justification of the project, the PWA and Ickes stressed the impact on relief to the region. Using the records of the Texas Relief Commission from late 1934, Ickes announced that within a fifty-mile radius of the Buchanan Dam there was a total of 23,997 relief cases on record and "the number of persons dependent on relief therein aggregated to 95,422." In the ten counties that were part of the Lower Colorado River Authority there were 8,429 documented relief cases out of a total population of 22,869. The funding agreement outlined strict guidelines for an estimated 4,400 new jobs and directed that "not less than 80% of the total cost of the project will be expended by July 1, 1936. Priority was to be given to local workers and local suppliers."[54]

The $20 million in funds for the LCRA was located in two separate accounts each designed for a separate purpose. An initial $5 million grant was to be drawn from the Bureau of Reclamation for promoting flood control. The remaining $15 million was a combination loan and grant to finance construction and to cover that portion of the project not designated as flood control. Guidelines of the PWA established $10.5 million of the latter amount as a loan repayable by the purchase of revenue bonds by the authority and subject to the approval and satisfaction of the administrator. The remaining $4.5 million was considered a grant. All funds were made immediately available.[55]

All Central Texas was elated, both about the new job possibilities and the prospect that, at last, a meaningful effort was at hand to harness the Colorado River. The combination of flood control, irrigation downstream, employment, a new reservoir for recreation, and the subtle implication of hydroelectric power promised a bright future.[56]

Buchanan in Washington and Wirtz in Austin had at last prevailed. Wirtz realized that the timing was right; delay at the local level would have set back work on the Colorado for years. Buchanan's warnings to Austin to move with haste to secure funding from Washington were well founded. By April, 1935, the PWA had committed itself to 19,004 projects in 3,040 of the 3,073 counties of the nation. Over two million jobs had been created on "make-work" construction sites nationwide. Of the $3.76 billion appropriated for recovery under the Federal Emergency Relief Act, approximately $2.57 billion went to PWA construction projects; the remaining $1.2 billion was divided among the Civil Works Administration, the Civilian Conservation Corps, the Tennessee Valley Authority, and Farm Credit programs.[57]

Harold Ickes estimated that by mid-1935 the full impact on the economy had resulted in at least ten million persons having jobs and other forms of relief. Ickes was proud to claim that millions had "benefited by expenditures to increase the national wealth through construction of *useful* public works as a substitute for direct relief."[58]

Central Texas had been fortunate in its persistence. However, the ultimate key to success was the close cooperation between Washington and Austin. There had been compromise at both national and local levels. Water rights, hydroelectric power, West Texas interests, selection of the site for the 1936 Texas centennial, controls on the funding from Washington, and the constant reminder that the Colorado River could once again flood if not regulated provided for a volatile "political whirligig" for which a solution was found. The time had now come to begin rehiring to complete the Buchanan Dam and plan additional multipurpose structures for the Colorado River.

3.
CONSTRUCTION AND CONFLICT: GROWING PAINS OF THE LOWER COLORADO RIVER AUTHORITY

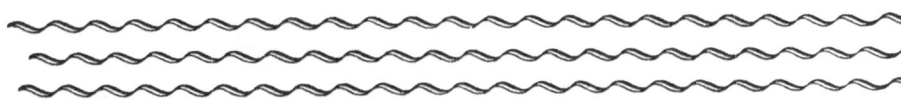

> *It is to such multiple-purpose projects as this that we are turning as a Nation in our fight to conserve our natural resources and thus preserve and maintain the civilization that we have built.*
>
> —Harold L. Ickes
> Secretary of the Interior
> Marshall Ford Dam Dedication
> February, 1937

Prior to the passage of the Lower Colorado River Authority Act, in November, 1934, little consideration was given to the large-scale construction of dams on the Colorado. The new authority assumed control of the equipment and assets from the Colorado River Company and C. G. Malott in February, 1935. The LCRA now had the task of completing and submitting federal applications as well as organizing a full complement of engineers, hydrologists, subcontractors, and workers.

Paper work and contracts to initiate the flow of funds from Washington to Austin were detailed and cumbersome. In addition to the guidelines of the LCRA Act and the restrictions imposed by the Public Works Administration, the authority had to comply with numerous employment, safety, and construction standards. The task before the LCRA was of a scale never before realized in Texas or the Southwest.[1]

Although the mechanics of obtaining the $20 million in federal funds were tedious and lengthy, the authority finally succeeded. The allotment to the LCRA was third only to the Hoover Dam project,

Major flooding along the Colorado River in 1935, 1936, and 1938 heightened the outcry for control of the river. This June 15, 1936, view of Congress Ave-

which received $38 million, and the Grand Coulee Dam project in Washington State, which was allocated $23 million.[2] In total, the federal government set aside $100 million for use by the Bureau of Reclamation for flood control and irrigation. The financial commitment to Texas was a substantial one. However, the LCRA did not have the luxury of time to organize fully. In addition to the problems caused by those who continued to question the validity of the authority, the project, within the first eighteen months, encountered a number of major time-consuming delays and problems. From 1935 through 1937 the project was delayed by untimely floods and managerial problems. It faced the potential of losing a substantial part of its federal allotment, continued to be harassed by the public utilities companies, and experienced bad public relations with local subcontractors. The most pressing problem, the one for which the authority was primarily created, was flooding.[3]

During the early construction period, prior to 1940, a number of floods served as further reminders of the primary purpose of the dams and reservoirs. The Colorado River was usually a very placid,

CONSTRUCTION AND CONFLICT 45

nue in Austin demonstrates the magnitude of flooding in Central Texas. (Courtesy Lower Colorado River Authority)

meandering stream. However, three floods in 1935, 1936, and 1938 had a tremendous impact on the evolution and final outcome of the project. The first of the three floods occurred in June, 1935, taking its toll in Austin and the lower Valley. In addition to the loss of life, over $16.5 million in property was destroyed.[4] This flood was the second largest on record; "the best records indicate that the rise of 1869 reached a stage from 1.5 ft. to 3 ft. higher than this rise in 1935."[5]

The 1935 flood wrecked the Austin Dam, carrying away most of the superstructure, the low-lift pump station, and the city water plant. Hundreds of houses and businesses were destroyed, "brick structures were taken out, entirely, not even the foundation being left!"[6] Consulting engineer Frank W. Chappell reported six-room houses floating down the river. In terms of the discharge, the flood would have placed a strain on any structure along the river:

> The stage of the river during the crest of this rise was reported as 41.7 ft. The State Board of Water Engineers is making an estimate

This dramatic picture captures the moment a house went over the spillway of the Austin Dam in the 1935 flood. (*Courtesy Lower Colorado River Authority*)

of discharge of the stream, but has not completed it. It will likely run over 400,000 sec. ft. The flood which wrecked the dam in 1900 had a discharge of 150,000 sec. ft.[7]

Two other destructive floods followed in 1936 and 1938, keeping the LCRA focused on flood control.

Not only flooding but also competition for control by various interest groups to determine who would manage the new LCRA created problems. Given the large-scale construction aspect of the project, the general manager would have substantial authority. Transition of assets from the Colorado River Company to the LCRA had, for the most part, gone smoothly. In an effort at some equity, private claims placed by both the Fegles Construction Company and the Fargo Engineers Company were honored. The actual disbursement of the money to the Colorado River Company is, to this day, a mystery; the records, however, indicate that a majority of the $2.2 million in settlement was divided among Ralph Morrison, C. G. Malott, and Alvin Wirtz. It was from this settlement and the transition to the new governing board of the LCRA, that the first major controversy emerged.[8]

As former president of the Colorado River Company, Malott contended, with some justification, that he should be the new general manager of the LCRA. He had acquired some expertise and knowledge of construction at Buchanan Dam, and was quietly supported for the job by his father-in-law, Ralph Morrison, as well as by Wirtz. Wirtz, acting as the local power broker, urged the board to accept Malott and even contacted Buchanan in Washington, D.C., in an effort to obtain a favorable endorsement. Nevertheless, the new board, chaired by Roy Fry, stood firm in its opposition to Malott, primarily because it was believed that his family ties to industrialist Morrison would make him sympathetic to private power interests. The board also considered Malott to be brash and authoritarian. A critical point was reached in late August, 1935, when the board members threatened to resign rather than appoint Malott. Wirtz's failure to have Malott appointed the first general manager proved to be very ironic; it was one of the few goals Wirtz had failed to achieve. The record shows that Malott did in fact recognize the engineering problems at Buchanan Dam as well as understand the financial needs of the entire multidam LCRA project. As early as February, 1935, he had already provided detailed engineering and financial data to the Bureau of Reclamation. His estimated cost to complete a unified dam and power system was $20 million, a figure that proved to be very close to all other estimates, excluding the Marshall Ford Dam. To resolve the issue, Buchanan urged Harold Ickes to select a suitable candidate.[9]

Ickes considered the selection with care and recommended, or offered to "lend," Clarence McDonough, chief engineer of the PWA, to be general manager for a period of two years.[10] McDonough had excellent credentials to supervise what the press was calling "one of the three major power-and-flood-control projects West of the Mississippi."[11] Representative Buchanan functioned as an intermediary between the LCRA board and the PWA, securing for McDonough an annual salary of $15,000 as specified by Ickes. Pleased with the turn of events, Buchanan telegraphed the authority chairman, Roy Fry:

> McDonough has had extensive experience in building dams, bridges, docks, massive tunnels, mining and industrial construction and reclamation works in this country, Canada, South America and Europe and has gained a worldwide reputation as an engineer of first class ability. He is admirably suited and will form a cooperative link between the Authority and public works administration and reclamation bureau, and his services will save a hundred times his salary.[12]

McDonough's experience and PWA connections were expected to prove very valuable.

Clarence McDonough arrived in Austin in September, 1935, to initiate operations of the new authority. Housed in rented offices in the Littlefield Building, McDonough organized a staff and analyzed the pending project. More than just his impressive engineering skills would be needed to maintain political equilibrium between Washington and Austin.

In order to outline areas of responsibility, McDonough held a number of meetings with representatives of the Bureau of Reclamation and Henry Hunt of the PWA. Initial plans called for the bureau to "perform all the [engineering] functions possible," with the authority giving administrative assistance in Texas.[13] The key question in all these discussions was who would control the $5 million allotment to the bureau and the $15 million loan/grant to the LCRA. In order to direct construction in Texas, the bureau entered into an agreement, by contract, with the LCRA. Once the bureau had exhausted its initial fund of $5 million to "construct the flood control and irrigation features of the project," an additional $5 million was to be made available to the bureau by the LCRA.[14]

The authority would be primarily concerned with local matters of rights-of-way, limited contracts, labor availability, and material purchases. That funds from multiple sources would be used concurrently and charged to various accounts was of some concern. Referring to Ickes's letter from R. T. Elliott of October 12, 1935, the acting comptroller general of the United States endeavored to clarify justification for the new project: "The project consists of a unified system and series of dams, the construction of which will be commenced at intervals and the work will progress concurrently, some of which, if not all will have flood control value."[15]

In mid-October, 1935, as crews prepared to start clearing land around the half-complete Buchanan Dam, bad news reached Austin. A reevaluation of projects nationwide—in large part due to a shortage of funds—had caused the Bureau of Reclamation, without warning, to withdraw $3 million from the Colorado River reclamation allotment of $5 million, leaving a balance of $2 million. The total allotment to the bureau by the Federal Emergency Relief Administration for nationwide projects had been cut by $20 million.[16] This administrative decision coincided curiously with the authority's approval to allow the bureau to use the additional $5 million of its loan/grant if needed. The federal comptroller general had on October 12, 1935, given official notice to both the secretary of interior and the Bureau of Reclamation as to the final interpretation of the status of the funds allocated to Texas:

> The $15,000,000 allotment is specifically for the purpose of financing that portion of the project "not provided for by the allocation of $5,000,000 to the Department of the Interior." It is thus apparent that

the $5,000,000 allocation to your department [Interior, i.e., Bureau of Reclamation] is first to be exhausted, and the remainder of the cost of the project is thus to be paid for from the $15,000,000 allotment.[17]

Reaction from Texas to the proposed reduction was both swift and blunt. The Colorado River project could not survive without its full allotment of $20 million guaranteed by the government. At the forefront of the protest was Congressman Buchanan. Protesting to Ickes, Buchanan reminded the secretary that "not one dollar more than was necessary was allowed in estimating the actual cost. The necessary cost for construction was itemized in great detail and these items added to the aggregated Twenty Million Dollars."[18] Buchanan was no stranger to the constant struggle over the development of the Colorado. Writing from his home in Brenham he questioned the legality of reducing the allotment and reminded the secretary that $5 million had been allotted to the Reclamation Bureau for the "specific purpose" of aiding in the flood control of the Colorado River in Texas.[19]

An enraged Buchanan went on the offensive. Realizing that the reduction of funds would jeopardize an expanded series of dams, he singled out the Bureau of Reclamation as responsible for shifting funds away from Texas. Given that Ickes had taken the matter under advisement, it became clear to Buchanan that another crucial test was before the authority. Using the letterhead of the "Lower Colorado River Authority – Austin, Texas," Buchanan hit at what he considered the root of the problem: mismanagement within the Bureau of Reclamation. He quickly got to the point. "Work on this project has not been expeditiously handled to date," he charged, "as the Bureau of Reclamation was *unable* to spare the services of their chief designers for sufficient time."[20]

As of November 1, 1935, no work had begun on the Hamilton (Buchanan) damsite. Representative Buchanan presented a detailed status report to Elwood Mead, outlining the entire project from Austin upstream past Arnold Dam, Marble Falls Dam, Austin Dam, and proposed Marshall Ford Dam. Irritated over the continued delays and pending loss of millions in funds to his pet project, Buchanan pointedly advised Mead: "As you may know, the Colorado River project has been treated as an orphan, while some other projects . . . have been pushed"; he added that "this project needs the three million dollars which was temporarily taken."[21]

Buchanan further informed Mead that "the Reclamation Services are under contract with the Authority to *spend* the *full* five million on flood prevention and also to spend this amount *in advance* of spending of the funds given to Reclamation by the Authority."[22] In response to Buchanan's detailed letter, Mead replied that the entire matter "is a great surprise to me. It is evident that you do not understand the situation which exists here."[23] Mead continued with a complete rejection of Buchanan's assessments, stating that he had just received the first engineering report on the project. He only confirmed Buchanan's claim that little had been done past the initial report phase. At no time did Mead point out that the FERA was to blame for initially cutting the money, not the Reclamation Bureau.

Mead and the bureau had underestimated the response from Texas. Clearly, McDonough had been consulted to strengthen Washington's position and Buchanan had not. However, the problem, as Mead confirmed, ran deeper than the mere allocation of funds. In his role as commissioner of reclamation, Mead informed Buchanan that "information has come to me [from his engineers in Denver and possibly from Malott, who had attempted to be a "consultant" with the Bureau of Reclamation][24] that the Colorado River Authority questions the extent of my authority."[25]

The power struggle between Austin, represented by Congressman Buchanan, and the Bureau of Reclamation's Commissioner Elwood Mead in Washington was intensifying. Mead's attempts on behalf of his agency to redirect funds away from Texas were hasty and poorly planned. However, the matter of the allocation underlay a more serious question: who would exercise authority over certain segments of the Colorado River project in Texas?

In an effort to resolve the problem, Mead called both the chief project engineer at the Buchanan Dam site, Howard P. Bunger, and McDonough to Washington for consultation.[26] As a result of these meetings, McDonough outlined a solution whereby the authority would continue to have all local on-site responsibility for such matters as payment for labor, purchase of materials, and administration in exchange for the bureau's having design and inspection authority. McDonough and Mead, now in basic agreement, submitted a preliminary report to Secretary Ickes in mid-November; however, the question of the $3 million reduction from the Colorado project was not yet settled.[27]

Both Mead and McDonough simultaneously contacted Ickes with regard to the progress of the Colorado River projects. Delays persisted and a portion of the funding was in question. The LCRA and the bureau were moving in separate directions. In a memorandum to Ickes, Mead conceded that "no definite determination of the limits of authority and responsibility of these two organizations has yet been made. That misunderstandings or conflicts should arise is almost inevitable; in fact, they have already appeared."[28] The complex set of circumstances was difficult to grasp in the long-distance exchange between Washington and Austin. It is clear that Austin overestimated the financial needs of the Colorado project and that Washington underestimated the magnitude of the local problems and politics in Texas. For example, in the ongoing investigation into Buchanan's claim that the project was delayed and mismanaged, Mead realized that the contracts between the bureau and the LCRA omitted a number of key considerations, such as design features to produce "power." Mead advised Ickes that "generating plants are so intimately connected with the construction of the dams, gates, outlets, etc., it is *infeasible* to design and build the dams without including the power plants."[29]

Having considered the problems and the objections of the LCRA, officials in Washington decided to force a settlement. McDonough and the LCRA board demanded only a fair review of what they were up against. After weeks of confusion and delay, Ickes ordered Mead to go to Texas in order to get firsthand knowledge; "such a visit also would promote *cooperation,*" the secretary noted.[30] Accompanied by William Kubach, chief accountant for the bureau, and Raymond F. Walter, chief engineer for the bureau, Mead departed Washington on December 11 for a week-long visit to Texas.[31]

The on-site inspection and meeting with the LCRA board was both beneficial and timely. The first full-scale federal investigation by Mead confirmed earlier speculations concerning the magnitude of the multipurpose Colorado dam project. "It is doubtful if all the work proposed can be built with the funds allotted," he concluded.[32] The investigation made clearer the constraints on federal-state cooperation on large projects. Although the role and scope of the Colorado River projects were understood in general terms, the actual specifications had not been clarified or completely agreed upon in Austin and Washington. The magnitude and multiplicity of PWA and bureau projects in 1935 limited the amount of close attention paid the LCRA project which might have ensured a smooth start. From the vantage point of the LCRA board and Buchanan, the Colorado project was their number-one priority; from the view in Washington the concerns of Central Texans ranked no higher than those of hundreds of thousands of other Americans in Nebraska, California, Colorado, or Arizona.

However, the magnitude of the bureau's programs did allow for cooperation, as suggested by Ickes. Additional political pressure from Buchanan hastened a solution. On January 3, 1936, shortly after the return of Mead from Texas, Ickes announced the full restoration of $3 million toward the construction of the Colorado River project. In a public statement, Ickes stressed the allotment was intended "to regulate the Colorado River of Texas for flood control, power generation and for supplementary water supply [irrigation] . . . below Austin."³³ This reversal ensured that the Colorado River project would get the full $20 million loan/grant approved by Roosevelt on May 28, 1935. To avoid any further delay or confusion the bureau assured the authority that it would move at full speed during early 1936.³⁴

The path appeared clear for the authority to proceed with haste to complete the still half-constructed Buchanan Dam as well as other structures on the Colorado. At least for the time being differences with Washington were resolved. On the authority's first anniversary the agency had over $11 million in construction under way or scheduled, and employed eight hundred workers, half of whom were involved in repairing and clearing the Buchanan damsite. The remaining four hundred employees were engaged in surveying between the Buchanan Dam and Austin for other potential damsites. They also began irrigation studies of the area between the capital and the mouth of the river. Engineers at the Bureau of Reclamation office in Denver were working on major designs for the dams. McDonough at last presided over actual work in progress in Texas rather than engaging in political maneuvering.³⁵

Clarence McDonough and the LCRA board had charted four major projects and a number of lesser ones for the control of the river. Completion of the dams had first priority. By mid-1936 the authority hoped to have concurrent site operations and construction in progress to finish Buchanan Dam, to complete the site selection for the Marshall Ford Dam, to accept bids for the smaller Roy Inks Dam three miles below the Buchanan site, and to work out an agreement with the city of Austin for selecting a new site for construction of a dam.³⁶

On the national scene the United States government accepted the completed Hoover Dam from the contractors on March 1, 1936. The project was completed two years, one month, and twenty-eight days ahead of schedule. Over 4.4 million cubic yards of concrete were used. The success of Hoover Dam greatly influenced similar projects throughout the western states.³⁷ In Texas, the Lower Colorado River Authority was receiving more attention than any other project in the state. The Red Bluff project on the Pecos River and the slightly older Brazos River Conservation and Reclamation District, which primarily involved the construction of a dam at Possum Kingdom, were both overshadowed by the LCRA.³⁸

After three years of the New Deal, the country had witnessed a tremendous change in the involvement of the federal government at the local and state levels. The government had committed itself to financing over forty-five water projects at points all across the nation. These projects were designed to create employment and to im-

A downriver view of the arch section buttresses 6 through 13 at Buchanan (Hamilton) Dam. *(Courtesy Lower Colorado River Authority)*

prove or correct an existing problem, such as flooding, navigation, or irrigation. Without grants and loans from Washington most state and local governments would not have had sufficient capital to undertake such large-scale projects. The loan portion of the federal allotments could be repaid either by letting bonds or by building projects that were self-liquidating. The prospect of selling hydroelectric power would enable the states to sell the requisite bonds, with the income pledged to repayment. Resenting the competition, the private utility companies challenged the legality of these government-sponsored power projects. The initial judicial review of the legality of such public projects centered around the TVA.[39]

The first challenge that could affect the LCRA came in September, 1934, when thirteen stockholders of the Alabama Power Company filed suit on the grounds that the federal government was illegally and unconstitutionally involved in the power industry through the TVA, primarily to defraud them of profits by means of unfair competition. This case, settled in February, 1936, resulted in an eight-to-one Supreme Court decision in favor of the TVA.[40] Chief Justice Charles Evans Hughes, writing for the majority, held that the construction of the Wilson Dam came under the commerce and general welfare clauses of the Constitution. Thus, being constitutional, the TVA could market hydroelectric power created by the dam. However, the court's decision was confined narrowly to the Wilson Dam, which had been authorized by the National Defense Act of 1916. Such a narrowly focused decision failed to define clearly the status of other federal-state contracts to build dams, most of which had the implied goal of flood control and the potential of generating hydroelectric power.[41]

The view of those involved in the LCRA was that the Supreme Court's decision in upholding the TVA meant that the LCRA and over twenty other Texas authorities would receive favorable treatment if challenged in court. The Texas utility companies, however, did not believe the decision applied to the LCRA. In early March, 1936, seven Texas private utility firms—Texas Power and Light, Public Service Company, Dallas Power and Light, Gulf States Utilities, Houston Power and Light, Texas Electric Service, and Texas Utilities—filed suit in federal court in Washington, D.C., to stop the flow of federal funds to the Colorado River and Brazos River programs. The suit, which produced a ten-day restraining order, was filed against twelve federal government officials, including Secretary of Interior and PWA Administrator Harold L. Ickes, Secretary of War George H. Dern, WPA Administrator Harry L. Hopkins, and Secretary of Treasury Henry Morgenthau.[42]

The suit against the federal government stated that Washington, not the Colorado or Brazos river authorities, was dictating the type of dams to be constructed and that "these hydroelectric dams, when completed would compete unfairly and illegally with the facilities which the companies have already provided near these two rivers."[43] Speaking for the seven utility companies, W. H. Thompson, vice president of Texas Power and Light, explained the terms on which the private utilities attempted to block with government funding any operations in Texas. The primary purpose of the suit he said, "is to prevent construction of hydroelectric power on the Colorado and Brazos Rivers."[44] Furthermore, the utility companies raised no objections to the allotment of the original $4.5 million to complete Buchanan Dam if the purpose was flood control and irrigation. But Secretary Ickes and PWA officials had instead "developed a $20,000,000 plan for a series of dams and reservoirs on the Colorado for the generation of power and transmission systems" to supply electricity throughout central Texas.[45] Although Thompson clearly admitted that he meant to entangle the two major Texas river programs in lengthy and delaying litigation, he stated in conclusion that the utilities suit was being filed in Washington "to protect the interest of our company and to try to preserve for the people of Texas the original plans for flood control."[46]

Heretofore, Wirtz, McDonough, Buchanan, and LCRA board, and officials in Washington had all quietly agreed not to promote the power question. Yet they all knew that, although flood control was the most immediate problem along the Colorado, the clear long-range objective was, as it is now, to generate electricity. From the outset, promoters had intended for the LCRA to be a multipurpose project; however, hydroelectric power was not openly promoted or publicized. To push the project to completion without a major court battle had been the initial goal of the LCRA board. In order to achieve this, the power question received only incidental mention.

In response to the power companies' charges, the Lower Colorado River Authority denied that the first priority of the project was to produce hydroelectric power. Buchanan, who was hospitalized for a minor illness, retorted that it appeared to him "those companies got their facts wrong," explaining that power development was only the third consideration.[47] The possibility of lengthy injunctions was grounds for grave concern, since over eight thousand jobs could be affected statewide. In a public-relations blitz the LCRA reminded Texans that the Texas Power and Light Company had extremely high rates, that only one farm out of fifty in the state had electricity, that the private utilities ran their companies for the good of the stockholders and not the customer, and that "no relief has been in sight for consumers except by the government yardstick!"[48] In the official LCRA statement, General Manager Clarence McDonough avoided the power question completely: "The dams of the Colorado program are being designed by the United States reclamation bureau, whose sole concern is flood control and irrigation, and are being planned as the most effective flood control units it is able to design."[49]

The extent of the initial ten-day restraining order was not at first fully understood. Wirtz, as legal counsel for the LCRA, interpreted the order to apply only to the activities of the Bureau of Reclamation and not to normal day-to-day activities of the authority. One point was clear: the true intentions of the utilities were at last revealed. The response from local citizens was one of outrage. Residents of Llano and Burnet counties did not agree with the position taken by A. J. Duncan of Fort Worth, president of Texas Electric Service, that power construction in conjunction with federally funded flood-control projects would "defeat the purpose of flood control, and make it impossible."[50] A Llano druggist replied, "It appears to me that the public utilities who have filed this injunction . . . are not willing to let the wheel of business recovery spin if it rolls over their toes; but instead are willing to let the working men pay, and pay, and pay."[51]

In fact, the restraining order had little effect on progress at the Buchanan Dam or on the preparation of the other damsites. Some supplies and one weekly payroll in Dallas from the U.S. Treasury were delayed. Representatives from the LCRA, led by Fry, Wirtz, and McDonough, joined Lewis Mims and John Norris of the Brazos River project in Washington on March 12 to review the problem with the Justice and Interior departments. Ickes did not seem overly alarmed, declaring that power was only "incidental," but privately expressed concern over the employment of thousands of workers not only in Texas but also nationwide.[52] On the same day the Texas representatives arrived in Washington, Ickes informed the Senate that thirty-six different injunction cases brought by utilities companies were pending.[53]

Within days the federal court agreed to a modification of the Texas restraining order. The court stipulated that no delay was to affect either payrolls or the purchase of materials.[54] A hearing on the order was postponed until late March. Buchanan claimed "the whole suit ought to be dismissed.... The idea of the utility companies placing a few dollars above human misery, destruction and loss of life caused by the kind of floods we have down there is inconceivable to me."[55] The final determination of the issue would be delayed for months and await a Supreme Court test case selected from over sixty similar suits filed against numerous New Deal power and reclamation programs.[56]

That Ickes characterized power on the Colorado as only "incidental" belied the fact that there were genuine concerns in both Austin and Washington over the possible response of the Supreme Court in future cases. At the instigation of Wirtz, Fry, and McDonough, Washington authorities began considering alternatives. All concerned recognized that lines of authority between the LCRA and the Bureau of Reclamation had to be drawn more clearly. In secret negotiations LCRA and bureau officials began looking for ways to shift as many contracts and responsibilities away from federal agencies and into the hands of local authority. As of late March, 1936, the LCRA felt that the initial allotments and construction agreements tied the authority to bureau guidelines. Furthermore, the LCRA felt they were victims of the whims of both the bureau and PWA over their funding. In the initial agreement part of the LCRA allotment of $20 million was a loan and grant, involving federal-state cooperation and classifying the project as not purely a federal one.[57] Of the bureau's grant of $5 million, only $209,028.96 had been expended and, of the $5 million advance to the bureau, only $6,748.91 had been spent as of March 20, 1936.[58] In a review of these expenditures by Kubach, accounting agent of the bureau, it was determined that "the unobligated balance could be readily returned to the Authority if the responsibility of the project is turned back to the Authority."[59]

The timing for such a move was greatly favored by the LCRA. Through the end of March, 1936, less than $2 million of the $20 million LCRA allotment had been spent. Contracts to date had been minor in nature—for constructing scaffolding, clearing around damsites and new reservoirs, and surveying. The capital-intensive phase involving the purchase of large amounts of steel and concrete was scheduled to begin in July. If the LCRA assumed full control then it alone could determine how the balance of $18 million would be spent in Texas. Moreover, the friction between the LCRA and the bureau would be eliminated or at least greatly reduced.

However, LCRA representatives, Wirtz in particular, did not want to sever all ties to Washington—only those involving the initial $20 million loan/grant allotment. Since the allotment was secure, Austin would not have to deal with such federal agencies as the FERA, PWA, the bureau, and the Department of Treasury. And finally, the transfer of funds and control would put some but not much distance between the LCRA and the Texas private-utilities suit against the federal government. Wirtz and Fry believed that if a separation could be achieved, the LCRA would have greater influence and control over the authority's projects, including the eventual generation of power. The transition of all rights and control to the LCRA would momentarily clean the slate. Careful not to break completely with Washington, Wirtz (knowing secretly that the separation of the LCRA and the bureau was pending) made the following statement at the Burnet Chamber of Commerce annual banquet:

> Whenever these utilities have tried to kick us in the face, the result has been that they kicked us upstairs. They attacked the four million five hundred thousand dollar original program; a twenty million dollar program was the result. I believe positive good to this project and this section will result from their present attack.[60]

When Wirtz made this comment, he knew that much more money and continued cooperation with the Bureau of Reclamation and the Department of Interior would be needed to finish the entire Colorado project. He supposed that Congressman Buchanan in his capacity as chairman of the House Appropriations Committee would meet any future needs.

Construction and preparation of both the Buchanan and Roy Inks sites continued to proceed without delay. In April the first contract for 365,000 barrels of cement at a cost of $835,159 was awarded to the Republic Portland Cement Company of San Antonio.[61]

The potential transfer of responsibilities did not seem to influence any of the day-to-day dealings between Washington and Austin. By mid-April Ickes was fully aware of the pending change and urged acting Commissioner of Reclamation John Page to investigate the matter with care prior to making a final decision. By early June rumors surfaced that a split with the bureau was under way that would place the LCRA in full charge. The primary concern for Commissioner Page in Washington was what the future relationship would be with the LCRA and how the bureau could honorably withdraw from Texas so as not to take the blame for any past or future problems at either the Buchanan or Roy Inks sites.[62]

Page's concerns were well founded. Reports from the Buchanan damsite engineer painted a bleak and confusing picture. During 1936 McDonough had steadily taken matters into his own hands without any consideration for bureau policy, and there were ill feelings. Site engineer H. P. Bunger outlined numerous problems. Upon Bunger's requesting two heavy-duty rock buckets for excavation, McDonough arranged for the purchase of two secondhand and largely useless sand buckets from the Middle Rio Grande Conservancy District. Attempts were continually made to use old water pumps, "a part of the junk leased from [the defunct] Fegles Company," for the water supply. Bunger reported the old pumps, frozen and with cracked and brazed casings, constantly ran hot. "I [Bunger] requested the purchase of a new pump. Mr. McDonough did not think this was necessary and delayed ordering one." The size of the Hamilton (Buchanan) damsite was scarcely considered. "As you know Hamilton Dam is located on both sides of the river. The camp and present water supply are on the south side. The dam on the north side is over a mile long and to supply water for this work we planned on pumping muddy river water to a settling basin and thence to an elevated unit on this side of the river as we had on the south side so they could be interchanged in case of emergency.... McDonough did not like our plan."[63] Problems of jurisdiction continued as McDonough, without authority, promised the aid of bureau engineers for downriver surveys. McDonough, as well as LCRA board members, attempted to decide who would be employed as well as their pay rate. Both the authority and the bureau attempted to determine policy for the site hospital, and the commissary contractor set up a gas and auto-repair station in the construction camp, "without so far as I can ascertain, bringing the question to our [the bureau's] attention."[64] The situation was obviously not conducive to productive work conditions on such a major project as the Buchanan Dam.

A final meeting, pivotal for the future of the LCRA, occurred on June 5. Present were E. K. Burlew, administrative assistant at the PWA; Congressman Buchanan, Roy Fry, Alvin Wirtz, Clarence McDonough, Commissioner Page, and a number of lawyers from the Solicitor General's office, the PWA, and the Department of Justice. All those present realized that more than the Buchanan Dam and the bureau's relationship with the LCRA were at stake. Buchanan dominated the meeting.[65] In effect he proposed a twofold solution. First, the bureau would guarantee to the authority the $5 million grant along with any other expenditures and would transfer all rights, contracts, and equipment for the Buchanan and Inks (also called the Arnold site) dams to the Lower Colorado River Authority. Page informed Denver that "the Bureau will ship to McDonough all the data pertaining to the Buchanan and Arnold Dams, and clean the office of all work in connection therewith."[66] This initial step put the LCRA in full charge of the Buchanan and Inks projects as well as the planned, reconstructed Austin Dam.

Second, up to this time, McDonough, Wirtz, and the LCRA had been brazen and contemptuous of the bureau and its operation at Buchanan Dam, yet they knew that they had neither the engineering expertise nor, more important, all the funding needed to build what was to be the major dam on the Colorado at Marshall Ford. This second part of Buchanan's solution, best described by Page, acknowledged the LCRA's shortcomings:

> He [Buchanan] desires the Bureau, however, to design and construct a dam at Marshall Ford or the Maxwell site. I think his reason is the necessity for additional funds for this dam and his feelings that these can best be obtained from Congress if the Bureau of Reclamation is building it. He expressed friendliness for the Bureau and recognized the unworkable arrangement which has heretofore existed, and instructed the Authority officers that hereafter, except for approval of the general plan for Marshall Ford, they *would have no contact* [my emphasis] whatever with its construction or design.[67]

Page advised Buchanan that to build a high dam of the size needed to replace the initial low dam design at Marshall Ford would cost more than $20 million. Only the $5 million returned from the bureau was available. A budget of $8 to $10 million would allow only enough funds to design and build a low-level dam. Buchanan wanted more than a low-level dam, so asked for one that could be added to for greater height.[68] Page objected and stressed the difficulty of a two-stage construction, but assured the congressman that all would be done to make best use of the initial funding in the hope that Congress would provide enough funds in the future to make it a high-level dam. (For a comparison of the low and high dam, see table 1 in chapter 4.)

Page and Buchanan knew a great deal more was at stake than the Colorado project. Prior to the June 5 meeting, Page had spent nearly two weeks with the congressman drafting a proposed appropriation measure for the bureau. Only an acting director, and not enjoying the power wielded by his predecessor, Elwood Mead, Page sensed a unique opportunity to expand the bureau's activities in the West.

While Buchanan was looking after the interests of Central Texas, Page was coolly pursuing every advantage for the bureau. In a confidential memo to Chief Engineer S. O. Harper in Denver, Page commented that he expected to have a budget to meet the needs of the bureau "for all projects, except Grand Coulee and Central Valley.... Besides being the best arrangement... there are manifest advantages" in cooperating with Buchanan.[69]

Appropriations for the reclamation fund would ensure a long list of new projects. Page stated confidentially, "I believe the assurance is given by the promise that Mr. Buchanan and the Administration will whole-heartedly and vigorously support this [Reclamation] program."[70] To guarantee Buchanan's support for the appropriation, Page advised the Denver bureau office to begin work immediately on designing the Marshall Ford Dam in such a way that as an initial low dam it could be raised to a high dam.[71] Removal of the Buchanan and Inks dams from the bureau's control was approved in mid-June by Secretary Ickes and officially announced on June 22, to be effective on July 1. Fortunately for the LCRA the pending court hearing filed by the Texas utilities was delayed until December, 1936.[72]

Bureau officials had other reasons to accomplish the transfer of authority as quickly as possible. Highly confidential correspondence exchanged by the bureau engineers in Denver and Austin suggested that there were problems at Buchanan Dam.[73] Bureau inspector Bunger reported that there was "considerable leakage developing through cracks in the arches when subject to a head water of only a few feet!"[74] The situation was critical. After only moderate rains in Llano and Burnet counties (no flooding) during late June, Bunger reported that three feet of head water had built up on the lowest ring or dam section downstream from the arches. This resulted in a somewhat strange phenomenon. The standing water was higher on the downstream side than on the upstream side. "Due to this difference of head [water on the downstream side was higher than on the upstream side], water is running from the downstream side to the upstream side in a steady flow through cracks which were noted last winter near the center of each arch."[75]

Never in the recorded history of the Colorado River had water run upstream. The bureau had a dilemma. In Austin, Bunger concluded in his report to Walter that reinforcing steel was exposed to water and rusting and that, "due to mud being carried into the cracks, I doubt if they will ever be tight even under load."[76] Walter in turn informed Washington that the engineers in the Denver office were of the strong opinion that their earlier stamp of approval (the Bureau of Reclamation seal) should not appear on any plans or drawings for the Buchanan

The September, 1936, flood raised major questions in Austin as to the goals and responsibility of the Lower Colorado River Authority. The state capitol is visible in the distance. *(Courtesy Lower Colorado River Authority)*

Dam project. He also recommended that all plans in the hands of the LCRA be recalled and that a letter be written to the authority making it clear that all documents prepared by the bureau were only preliminary in nature and subject to numerous revisions and detail modifications from time to time. In short, the bureau must at once "disclaim all responsibility for the adequacy of the plans or for the safety of the dam."[77] Furthermore, Walter concluded:

> Such a condition [cracks] does not lend a feeling of confidence in the integrity of the present structure and further emphasizes the advisability of having the Bureau relieved from all responsibility for its design and construction if the operations are turned over to the Authority. . . . It appears certain that those responsible . . . have no conception of the design and construction problems involved in the completion of Hamilton [Buchanan] Dam.[78]

Page expressed no alarm. Although concerned with the situation outlined in Walter's report, he seemed, as did the Denver engineers, more concerned with the reputation of the bureau. Page claimed that in his conference with LCRA representatives he had given fair warning: "I stated plainly before all of them, including Mr. Buchanan, that the Bureau was jealous of its reputation, that this was a dangerous river and that the designs for both of these structures were incomplete, and stressed especially the fact that the Authority organization was totally inexperienced in this type of work."[79] Accordingly, he advised Walter to draft a letter to this effect in order to absolve the bureau from responsibility for defects in the dam.[80] No record exists as to whether the LCRA was contacted on the issue of the cracks.[81] Heavy flooding persisted during early July in South Texas along the lower reaches of the Colorado, San Antonio, and Guadalupe rivers.[82] Rain in the watershed north of Buchanan Dam remained light, posing no immediate threat to the dam.

To take advantage of the summer months and longer hours of daylight, the LCRA accelerated the pace of construction at the Buchanan and Inks damsites. On July 5, 1936, the first concrete, since Insull abandoned the project in 1932, was poured at the Buchanan Dam. Thereafter, numerous contracts for construction and materials were let, and work went forward simultaneously on both projects.[83]

At last construction was going forward at breakneck speed. Between 1,660 and 2,100 cubic yards of concrete were being poured daily. When Buchanan visited the sites in mid-July, there were 800 workers employed; by the end of the month there were 1,050; by late November there were nearly 2,000. The combined weekly payroll exceeded $60,000 for an average workweek of forty hours.[84]

As dam construction progressed in Texas, the maneuvering continued in the federal district court in Washington, D.C., resulting in four modifications to the original injunction. The utility companies argued that not only the project in Texas but also others nationwide should be delayed until a final ruling was obtained. In addition to the LCRA, sixty-one projects in twenty-three states, with an estimated total construction cost of $147 million from PWA loan and grant programs, were at stake. In the meantime the LCRA, encouraged by both Buchanan and Wirtz, worked as rapidly as possible to have as much of the Buchanan and Inks dams completed. The delaying action by Wirtz in Washington was designed primarily to allow the project to move forward. Not surprisingly, both the PWA and the Bureau of Reclamation were quietly supportive.[85]

By late 1936 Buchanan Dam had obtained both size and shape. Plans were rapidly being developed to build a temporary forty-foot-high cofferdam upstream in order to divert the river so that the last gap in the dam could be closed. The massive 1,960-foot-long dam standing 145 feet above the streambed was a dream come true for Congressman Buchanan. The size of the project, along with the prospect

The LCRA combined large-scale engineering as well as a major New Deal employment source in Texas. Workers at Inks Dam pose in front of the main thirty-foot-diameter discharge penstock, April, 1937. (Courtesy Lower Colorado River Authority)

of more dams and more funding, struck him as he stood on the spillway atop 65,000 tons of concrete and steel.[86]

After meeting with LCRA officials and visiting the damsites for the final time in late December, Buchanan returned to Washington for what he hoped would be the final court hearing and for the convening of the Seventy-fifth Congress. He arrived in Washington in time to find out that the Texas utilities' court hearing had again been delayed into early 1937, primarily because the U.S. Supreme Court was expected to hand down a ruling on another case that would include the Texas suit. The Court had agreed to hear the South Carolina case of Duke Power et al. v. Greenwood County et al. This case emerged as the test which would have dramatic bearing on over ninety other cases against the PWA.[87]

The Duke case against the PWA provided the best opportunity to end the long struggle over the issue of federal funding of public power projects. If the government won, all pending suits in Texas would be thrown out. If Duke won, Wirtz planned to pursue the Texas suit, calling for another court hearing to decide the Colorado River allotment on its own merits. The South Carolina case, it was expected, would prevail on the side of the public. This test case involved no flood control, irrigation districts, or any other water-use feature except the production of hydroelectric power and the legality of a funded public agency in competition with a private utility for customers.[88]

However, the LCRA could not afford to await the Court's deci-

An upstream view looking north along Buchanan Dam, here nearing completion. A special rail line was used to deliver supplies to the site, January,

sion. The Buchanan Dam project was reaching a critical juncture. The dam was nearly complete and construction had reached the point where generating turbines had to be installed. Wirtz petitioned the court to allow the turbines and generator work to proceed. The court agreed, and Wirtz contacted LCRA's chairman, Roy Fry, to let the contract at once.[89]

With the new year came greater promise for the continuation of LCRA projects. Threats from the power companies were dashed when the Supreme Court handed down a decision in favor of the government. The Court declared that the utilities "had no stand to question the validity of the proposed loan and grants because the competition of the municipalities (or authorities) was legal."[90] Numerous New Deal works projects received a boost. The timing for the LCRA could not have been better; the authority and all of Central Texas

1937. *(Courtesy Lower Colorado River Authority)*

realized that the dams and cheap power were soon to be reality.[91]

The outcome of the Duke case made the separation of the LCRA from the Bureau of Reclamation even more fortuitous for Texans. Even the bureau and the Department of Interior were surprised and impressed that the authority could advance the project at such speed.[92] The prospect that flooding might finally be controlled and that cheap, plentiful power would be available appealed to farmers and urban dwellers alike. Downstream from the nearly complete Buchanan and Inks dams, the construction of Marshall Ford Dam, eighteen miles northwest of Austin, held the key for a prosperous 1937 and 1938. Even the prospects for continued funding seemed bright. No one doubted that Congressman Buchanan, as chairman of the House Appropriations Committee, would uphold his pledge when Congress reconvened in January, 1937.

4.
MARSHALL FORD DAM:
THE 1938 FLOOD

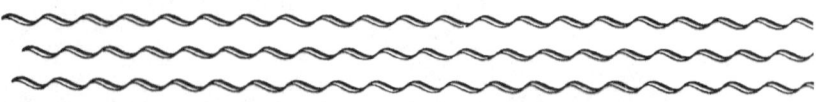

> *Down in Austin the problem of my friend, Congressman Lyndon Johnson, is to keep his land from washing away—washing down the rivers and into the sea.*
> —Franklin D. Roosevelt
> Amarillo, Texas
> July 11, 1938

The development of multipurpose dams on the Colorado River in Texas had become reality by early 1937. There was, however, a tedious, ongoing political process in Washington as well as Austin to sustain the Lower Colorado River Authority project. As a result of a change in Texas' representation in Washington and a jurisdictional squabble between the LCRA and the Bureau of Reclamation, a complicated series of events transpired that determined the fate of the Marshall Ford Dam.

With Alvin Wirtz in Austin and Congressman Buchanan in Washington, the state and the LCRA had an effective team working to ensure the success of the vast multipurpose river project. Buchanan as chairman of the House Appropriations Committee was in a powerful position to shepherd his pet projects. The opening of the Seventy-fifth Congress in early 1937 held the promise of approving the final appropriations needed to ensure funding of the linchpin of the LCRA network, the Marshall Ford Dam located eighteen miles upriver from Austin. All Central Texas was optimistic at the prospects for funding, since the building program was well under way. Such outward confidence was dashed on February 22, 1937, when Buchanan, at age seventy, died suddenly of a heart attack.[1]

Congressman Buchanan had contributed much to the growth of the LCRA project. The project's initial dams, Buchanan and Roy Inks, were fully funded and nearly completed. The second set of dams included Marshall Ford and a reconstruction of the previously destroyed Austin Dam in the heart of the state capital. This network of LCRA dams gained national recognition. The difficulties associated with Buchanan Dam had been in the limelight for nearly a decade. When finally completed it would be the world's largest masonry arch dam, confining over twenty-three thousand acres of water. It was hoped that the delays and intrigue experienced in completing the Buchanan Dam could be avoided at Marshall Ford. When complete, the Marshall Ford Dam would be the fifth largest dam in the world. A great deal was thus at stake.[2]

Prior to his untimely death, Congressman Buchanan had worked out an agreement between the Bureau of Reclamation and the LCRA, which described the role and responsibility of each agency with regard to the Marshall Ford project.[3] Ever mindful of a possible shortfall of funds, Buchanan, Wirtz and McDonough urged a prompt site selection and beginning of construction. In urging immediate action Buchanan wrote the PWA in Washington: "I am intensely interested in seeing that this *dam be started.*"[4] His impatience was well founded. Steps were initiated by the LCRA in late August, 1936, to speed the contract bidding process for the Marshall Ford Dam, estimated to cost $10 million under a grant from the Bureau of Reclamation.[5]

Enough funds were available under an initial $5 million appropriation to the LCRA to start construction, and bids were requested. By October 20, 1936, three bids were submitted: $5,781,000 jointly by Brown and Root, Inc., of Austin and McKenzie Construction Company of San Antonio; $5,909,000 by Utah Construction Company of Salt Lake City; and $7,322,000 by W. E. Callahan Construction Company of Dallas. After noting each proposal, the bureau office in Austin forwarded the three bids to Washington for final consideration and approval. In early December, Brown and Root, Inc., and McKenzie Construction were awarded the contract to construct a two-stage "straight-gravity type" dam with a final estimated height of 265 feet. The initial agreement thus stressed the need to build the dam in two stages. The first stage, or low dam, would be 190 feet high and would create a reservoir with a capacity of about 600,000 acre-feet, with the final high dam increasing storage capacity to over 3 million acre-feet.[6] The Marshall Ford contract involved one of the biggest reclamation projects west of the Mississippi River. Building the massive dam was a tremendous challenge to the new contractors.

George and Herman Brown, the principal owners of Brown and

Root, Inc., were primarily road builders. They had earlier completed reservoir clearing work at the Buchanan site but had never ventured into actual dam construction. Yet, as local contractors they were aware of the magnitude of the Marshall Ford project. Although novices in dam construction, their company was to serve as the lead contractor. They obtained engineering help from the bureau, since the construction and quality control were ultimately under federal authority. In addition, Brown and Root had ample local political connections to facilitate cooperation and to handle bureaucratic problems. The legal counsel for Brown and Root, Inc., Alvin J. Wirtz, was well paid for untangling snarls and writing contracts that were in favor of his clients. Wirtz, it will be recalled, was also legal counsel for the LCRA and the court-appointed receiver for the old Insull interests. That a probable conflict of interest existed was not acknowledged until years later.[7]

In the case of Marshall Ford Dam, the bureau was contracted to engineer the dam for the LCRA in cooperation with the local contractor. The intention was to avoid the jurisdictional problems experienced at the Buchanan Dam. Congressman Buchanan, in his earlier meetings with Commissioner Page, had made assurances that the LCRA would be the ultimate owner, yet design and supervising construction of the new dam rested with the federal government engineers. As the Brown and Root company laid plans to secure both the subcontractors and labor, the bureau in early 1937 announced that the chief government representative and construction superintendent would be E. M. Whipple, who would leave a similar job at the Chickamauga Dam of the Tennessee Valley Authority.[8] Whipple was only one of many professional engineers transferred to the LCRA project. On the same day that Whipple arrived in Texas, Secretary Ickes announced the award of a contract for $43,356 to build a "government camp at Marshall Ford . . . within 150 calendar days."[9]

During early February Brown and Root began preliminary site work. Since the dam would be placed in an out-of-the-way stretch of river, an eleven-mile railroad spur was needed. Roads had to be improved and widened, the camp completed, field headquarters set up, a cableway with two towers constructed, and heavy equipment secured. In all, Brown and Root invested $1.5 million in equipment prior to beginning construction. In order to get maximum exposure for the new dam project, Herman Brown and the LCRA announced plans for a groundbreaking ceremony in late February, conveniently to coincide with Interior Secretary Ickes's visit to San Antonio.[10]

For a number of months, Congressman Buchanan had urged Secretary Ickes to visit Texas and the Southwest. No other sitting Secretary of Interior had visited Texas; a visit would thus be appropriate, since for four years Ickes had played a major role in developing the Colorado River. The LCRA was seen as one of the primary reclamation projects as well as a possible model for other multipurpose western developments. Ickes, therefore, had a twofold purpose in visiting Texas. He chose San Antonio as the place to restate and defend the programs of the PWA.

By early 1937 the FERA and the PWA had expended over $2.5 billion of the ultimate approximate outlay of $4.1 billion. In order to justify the national PWA program, Ickes addressed the Associated General Contractors of America at their annual meeting on February 17. He boasted that nearly twenty thousand projects were under way nationwide, using over twenty-eight thousand principal contractors, and employing hundreds of thousands of Americans. Defending the government's public works program, which had come under attack as being unconstitutional, Ickes reminded the audience that to relieve unemployment "President Roosevelt had the imagination and the will to . . . set up our public works program." Furthermore, Ickes endorsed a plan for an ongoing permanent PWA program, stating, "We believe that what we have done is constitutional—in any event it is something of which we can be proud" and if a permanent department is established it will have the "plans, procedures and methods . . . to use *if* another depression should hit us!"[11]

After defending the national PWA policy in San Antonio, Ickes made plans to preside over the dedication of the Marshall Ford Dam and address a joint session of the Texas legislature. In both instances, he received a warm welcome. Local papers ran half-page stories on his visit, which stressed the impact and influence he had had with the Colorado River project.[12] With Ickes slated to detonate a ceremonial charge at the dedication of the dam, the *Austin Statesman* predicted that the secretary "will touch off a dynamite blast . . . whose reverberations in the cedar hills will echo fulfillment of a half century dream of Texas."[13] Ickes praised those present for having the "good sense to retain [Buchanan] as their representative." Without his "interest and energy," the project would not have been possible.[14] Again Ickes took the opportunity in his Marshall Ford speech to stress that the president's interest in reclamation was "more than academic":

> Every section and sub-section of the United States has one or more serious water problems. For too long a time these have been considered from a regional standpoint and in the light of an emergency situation. A national approach has been needed. A polluted stream in Massachusetts, a flood in the Colorado River of Texas, the exhaustion of the underground water in the California Valley, or a dust storm in the great plains area, all vitally affect people in every part in the United States.

It is for the common good of the country that, as a people, we are coming more and more to recognize that we have a common destiny and that we must advance toward it with a united front.[15]

Furthermore, construction of the Marshall Ford Dam, he concluded, "will constitute a service to Texas of National importance." Conspicuously absent from the colorful ceremonies was Buchanan. Although invited, he had stayed in Washington to tend to his pressing congressional schedule. Three days after Ickes's dedication speech at the Marshall Ford Dam, Buchanan died.[16]

The news shocked Central Texas. Alvin Wirtz, Tom Miller, Roy Fry, and others had all been active in efforts to protect the interests of the LCRA, yet it was Buchanan who had initiated the critical flow of federal funds to Texas. He had constantly tried to persuade the PWA to reconsider the Colorado River project, personally appealed to the president, and even appeared before the Texas legislature to ensure passage of the initial LCRA Act in 1933. His crucial and timely support had made the project a reality. Now, with his passing, two questions were raised. What would happen to funding for the LCRA and who would fill Buchanan's congressional seat on behalf of the Tenth District?

The answer to both questions came a few days later, on Sunday afternoon, February 28. The *Austin American*, in a bold headline, declared, "LCRA Project to be Carried to Completion Despite Loss of Buchanan—Colorado River Will Be Harnessed as Fast as Possible." The Texas delegation in Congress let it be known that the Colorado project would be finished. The second concern was answered by the young but ambitious state director of the National Youth Administration—Lyndon Baines Johnson.[17]

Twenty-eight years old, Lyndon Johnson had never run for public office. He had worked in Washington in the mid-1930s as secretary to Congressman Richard Kleberg and returned to Austin in late 1936 to direct the state National Youth Administration (NYA) office. Known to be both hardworking and outgoing, he found himself in a field of eight candidates. Thus, a solid campaign plan was needed to distinguish him. Alvin Wirtz, who had befriended Johnson some years earlier and referred to him as "m'boy Lyndon," directed the campaign, emphasizing the theme that Johnson was "Roosevelt's man" in Texas. Since all the candidates were pro-Roosevelt, Johnson, according to Wirtz, had to be "more" pro-Roosevelt. From the time of his first major campaign speech in early March at Southwest Texas State Teachers College in San Marcos, Johnson had been an avid Roosevelt supporter. He endorsed Roosevelt's controversial "court-packing" plan

Alvin J. Wirtz and Lyndon B. Johnson were lifelong friends. Wirtz served as Johnson's political mentor and confidant until his death in 1951. (*Courtesy LBJ Library*)

and quickly related it to the needs of Central Texas and the Tenth District:

> What good can come of having plans for controlling a Colorado River, Brazos River, or the Blanco and Guadalupe River, if in the end the Supreme Court shall say that these plans are not within the province of the government and the people themselves?
>
> Dams are being built subject to the order of a federal court in the District of Columbia. If the case is ever tried, decision will likely be appealed to the Supreme Court of the United States. Then this high tribunal whose members are governed as Mr. Justice Brandis once said, by their "previous economic predilections" will have the final word.[18]

Johnson followed the advice of Wirtz and won the congressional seat in early April.[19] In 1937 he was the youngest member of the Texas delegation.

Shortly after the special congressional election, Johnson was able to meet the man on whose coattails he had won. After concluding a fishing trip off Galveston, President Roosevelt came ashore to meet with Texas Gov. James V. Allred and the new congressman. The president invited Johnson to join the official party and ride the presidential train across Texas to Fort Worth, with a stopover visit at Texas A&M in College Station. The meeting and conversation were to prove useful in the years to follow. Roosevelt urged Johnson upon his return to the capital to contact his aide, Tom Corcoran, for future assistance.

Johnson's election to Congress placed him among a distinguished group of Texas politicians. He learned quickly the importance of such political contacts and depended heavily on Congressmen Sam Rayburn and Joseph J. Mansfield as well as Senators Morris Sheppard and Tom Connally, who between them had been in Washington over fifty years. New political alliances were thus formed as Wirtz, McDonough, and Fritz Engelhard (chairing the LCRA) in Austin and Johnson, Mansfield, Connally, and Rayburn in Washington cooperated to ensure maximum support for the dam projects in Central Texas.[20]

Concerned with the shortfall of funding for the LCRA from Washington, Mansfield, more so than Johnson, stepped in to deal with Commissioner Page at the Bureau of Reclamation as well as Ickes and his staff at the Department of Interior. The first priority was to secure the additional $5.5 million that Congressman Buchanan had planned to negotiate for prior to his death. Support for this budget item seemed secure since an independent report by the National Resources Committee commended Buchanan for his "economy program" that had requested only $5.5 million for completion of Marshall Ford. The report strongly recommended that at least $8 million be provided. This

Newly elected congressman Lyndon Johnson greets Pres. Franklin D. Roosevelt at Galveston in May, 1937. *(Courtesy FDR Library)*

amount, of course, was to be added to the $20 million already given the authority in 1935 and 1936.[21] As a seasoned legislator of twenty years and Chairman of the Rivers and Harbors Committee, Mansfield realized that an appropriation of $8 million would be difficult to obtain and that a smaller amount would be sufficient. He persisted, however, warning that a reduction in spending could affect the extension of existing projects.

As debate over more funds stretched into the spring, the Brown and Root company accelerated its activities to complete the site preparations at Marshall Ford. In 1937 the dam was considered to be totally a flood-control project; no provisions had been made to generate hydroelectric power. The Austin city fathers, however, realizing a future potential, agreed to erect electrical transmission lines from the city to the dam. The new lines could offer "an opportunity to get standby power from the hydroelectric plant at Marshall Dam *in the event* the LCRA *subsequently* [my emphasis] decides to generate and sell power."[22] In the meantime the city would earn over $150,000 by providing light and power to the plant over a two-year period.

In Washington the mood remained upbeat. Although under a federal restraining order to deny funding to generate power at either the Marshall Ford or Austin damsites, the Interior Department released a lengthy press announcement extolling the progress on the Colorado. The report claimed that, when taken as a total project, the work on all four dams was 40 percent complete.[23] This news played conveniently into Mansfield's hands: he requested and received a detailed assessment from the bureau on the cost of the flood-control plans for Marshall Ford. Page at the bureau responded that in order to provide "adequate flood control benefits and also supply water for irrigation and power, [Marshall Ford Dam] should be built to a height of 265 feet, instead of 190 feet as now proposed."[24] However, Mansfield could take little comfort in the director's pronouncement since Page had on more than one occasion changed his position on the need for a high dam.

To continue such a program of construction "without interruption" would require an additional appropriation of $5 million for fiscal year 1938. The debate over whether Marshall Ford would be a low or a high dam persisted until 1939. The benefits to be derived from high dams were compelling (see Table 1).[25] For under twice the estimated cost of the low dam, the high dam would have the capacity for nearly three times as much flood storage along with over four times as much reservoir storage. The estimates did not include the additional funding needed to generate power.

But, the dam height was only a persistent annoyance as Mansfield worked to draft the enabling legislation. Meanwhile, Johnson

TABLE 1. **Marshall Ford Dam Estimates, 1937**

	Height (ft)	Flood Storage (acre-ft)	Total Reservoir Storage (acre-ft)	Estimated Cost (Exclusive of Power Plant and Reservoir)
Low dam	190	710,000	710,000	$11,478,660
High dam	265	1,850,000	3,060,000	20,534,502

maintained a steady public relations campaign to keep key Washington agencies informed and his constituents at home assured of his commitment.[26] When Mansfield concluded his efforts to fund Marshall Ford, Johnson became involved with negotiations between the city of Austin and the LCRA over rebuilding the Austin Dam. Dealing with local Texas politicians proved more challenging than anticipated for the new member of Congress. Austin's leaders looked upon the Austin Dam as the property of the city. They wanted the benefits of the development without surrendering any property ownership or rights. Thus, the Austin Dam involved nearly as much controversy as the Buchanan Dam and Marshall Ford.

Memories of the dam's failure in 1900 were revived yearly. During his successful race for Congress, Johnson was reminded by a feature story in the *Austin American* that his duties would include harnessing the destructive power of the Colorado.[27] Such reminders were instrumental in calling attention to the possible rebuilding of the massive Austin Dam less than two miles from the steps of the state capitol. On numerous occasions the city of Austin had made efforts to improve the dam. Not surprisingly, by the mid-1930s the city council was prepared to reach out to either the LCRA in Texas or the PWA in Washington for funds to rebuild the dam. For some time Austin mayor Tom Miller and LCRA general manager McDonough had discussed the possibility of the LCRA completing the dam and making the necessary improvements to produce hydroelectric power, primarily for Austin. If an agreement could be reached, the dam would be "Unit No. 4" in the overall LCRA system.[28]

Negotiations for completion of the dam started in mid-April, 1937. A number of organizations questioned the legality of the Austin project. Objections were raised by Texas Power and Light, a former lessee of the site that had pledged yet failed to complete the dam. The Bureau of Reclamation expressed concern over the limestone foundation of the dam. Remembering the post-1900 flood report prepared by engineer T. U. Taylor, the bureau took a hands-off attitude toward the Austin Dam.[29]

Mayor Miller, who had already lobbied directly to the PWA in Washington, declared that the city had a number of options. The LCRA could be persuaded to complete the improvements for the city; Austin could follow through with the mayor's earlier efforts by again applying to the PWA for separate funding, or the city could finance the construction with revenue bonds.[30] Miller was disappointed because the LCRA had not taken a more active interest in the Austin Dam, and his appeals to Washington were designed in part to spur the authority to action. But, because of the injunctions pending in the fed-

To prevent water from undermining the Austin Dam, a phenomenon that twice destroyed it, crews in 1939 work to sink the dam's foundation below the limestone bedrock. (*Courtesy Lower Colorado River Authority*)

eral courts, the LCRA was reluctant at first to commit to the project, which included hydroelectric improvements. The Austin city fathers who had had sole ownership of the dam since 1931 believed they could complete the dam as a separate local project without being affected by any federal ruling. Moreover, the city did not pose the same threat to private utilities as did the larger, more extensive, multicounty LCRA. The damsite, Miller concluded, was the property of the city. Citizen reaction was voiced by advisory committee member A. J. Eilers, who supported the mayor: "God gave us that water out there and no man or set of men are going to keep us from using that water" to produce power.[31]

Faced with the harsh reality of needing a source for construction funding, the city struck out on its own to attract the needed money. A bond election could prove unpopular during such tight economic times; so, the city turned to Washington and negotiated with the PWA. In order to attract the approval of the president, the city council forwarded a recommendation to Washington that the yet-to-be completed Marshall Ford Dam be renamed the Texas-Roosevelt Dam in honor of the president.[32] When their ploys did not produce an offer from the PWA, the city fathers turned to the LCRA.[33]

The LCRA reconsidered its earlier hesitant position and offered to construct a new dam and powerhouse for $1.6 million, provided the city deeded the dam and all adjacent properties to the authority. In return, the city would be allowed to buy power at "substantially TVA rates while the authority is retiring the cost of constructing the dam."[34] The city council, with strong support from the Citizens' Advisory Committee, rejected the offer.

Following the council's reaction, a week-long debate raged among LCRA representatives McDonough and Wirtz and the city council, the citizens committee, the mayor, and the local press. The key objection that blocked a settlement was LCRA's insistence on buying the Austin damsite. The authority had no other alternative since it was bound by agreements with the PWA, which required ownership of projects under its control. After further delays the council again instructed Miller to circumvent the LCRA and appeal directly to Ickes as head of the PWA, Page as chief of the Bureau of Reclamation, and Johnson as a member of Congress. Johnson, who had remained in the background, was now compelled to act.[35]

Johnson arranged several meetings for Miller and even supported Austin's efforts to retain city ownership. After extensive conversations between Miller, Johnson, and Ickes, it was announced in late May that Johnson had "persuaded the public works administration to favor a lease plan" instead of selling the dam to the authority.[36] The proposal

Work crews pause in June, 1937, for a pictorial of Buchanan Dam. The tower in the center supported a giant cableway that delivered supplies and concrete

included a thirty-year lease to the LCRA, which would be obliged to build a dam sixty-three feet high and a power plant. The city would have full control of the lake created by the dam as well as distribute the power within a ten-mile radius of Austin. In a dramatic change of events, both the LCRA board and the Austin City Council approved the new agreement in early June. All parties praised the "splendid cooperation from Congressman Johnson."[37]

A milder than normal spring and early summer facilitated the massive construction efforts along the Colorado River. But, however bright the forecast was in Texas, an ongoing battle ensued in Washington over adequate funding for Marshall Ford.[38] More was at stake than the lease agreement for the Austin Dam or the rising water at Buchanan. The recent fight between the LCRA and the city of Aus-

to the center sections. (*Courtesy Lower Colorado River Authority*)

tin, along with the ongoing debate over the merits of an expensive high dam versus a low dam at Marshall Ford, caused the PWA and the Reclamation Bureau to reconsider the development of the entire lower Colorado River project. In mid-1937 two very extensive independent reports were drafted by bureau engineers who sought to determine the usefulness of the Marshall Ford Dam, its overall relationship to flood control on the river and the need for a high dam.[39] The timing of the reports was crucial.

Initially, the reports seemed destined for review and filing since Congressman Mansfield was able to push through the House his Rivers and Harbors Committee enabling act, which included an amendment providing $5 million for Marshall Ford. In the Senate, Tom Connally supported the Marshall Ford project as well as the Brazos River project at Possum Kingdom by earmarking funding in the 1937 relief bill.[40] Prior to the final passage of the funding bill for the Marshall Ford Dam, however, Lyndon Johnson in a letter marked "personal" advised Wirtz that "things seem to be confused here. . . . The attitude prevailing at the PWA is very definitely unfavorable to the [LCRA] project."[41]

Johnson had good reason to be concerned. The New Deal was under direct attack by those who questioned the massive spending programs. Although reelected in 1936, the Roosevelt administration stood at a crossroads in 1937–38. Roosevelt had been very successful in his dealings with Congress until he failed in his efforts to expand the Supreme Court.[42] Dozens of suits against PWA projects were pending in federal courts. At the same time, business interests questioned the legality of expanding federal government agencies into industries that heretofore had been private. Those who argued in favor of the expanded role of government watched in dismay as the nation descended into a severe recession in late 1937.[43] A slow-developing yet consistent faith that New Deal programs would achieve a recovery was dashed as unemployment skyrocketed and every major economic indicator plummeted.[44] In Congress a new coalition of conservatives, with representation from both parties, reflected the mood of the nation in questioning the validity of the New Deal programs.

To be sure, the Roosevelt administration was engaged in reevaluating policies and programs. At the center of these debates was puzzlement over the causes of the recession of 1937. Had the New Deal gone far enough—or too far? Clearly, Congressman Johnson had hopes that it had just begun to have a positive effect on the national economy. Yet he feared that as the administration reassessed the government's role in promoting recovery, it might take action that would reduce tangible benefits to his congressional district. Initially, at least, his worst fears were substantiated.[45]

In their efforts to bring the question of the Austin Dam directly to Secretary of Interior Ickes, Miller and McDonough unwittingly initiated a close scrutiny of the LCRA. The list of complaints and "serious questions" relayed by Johnson to Wirtz in Austin were representative examples of a growing negative reaction in Washington. Johnson outlined over a half-dozen problems and inconsistencies that placed the Colorado project in an unfavorable light. "They [the PWA staff and the secretary's main representatives] say" that there is a serious question of whether the authority has "enough money . . . to do the job." In order to shore up the lack of confidence, Johnson warned Wirtz that the authority must at once submit a revised and detailed budget for the Austin Dam, conduct a survey of current power usage, and provide "some definite data" on the disposal of the power to be generated. Johnson also noted that Ickes's staff expressed strong reservations about the estimates made by the authority, suggesting that these were "far below the actual cost." Most notably the LCRA was $1 million over budget on the Buchanan Dam. Further, the PWA lacked confidence in the estimates for the Austin Dam, which "will cost at least a million more" than proposed. Johnson assured Wirtz that "I am informed, the Administrator, Mr. Ickes, does not approve of any further steps to obtain money for Marshall Ford and will not aid in this connection until all of these things are settled to his satisfaction." Johnson concluded by requesting that Wirtz and McDonough come to Washington to brief the appropriate officials. Sensing the gravity of this situation, Johnson told Wirtz, "Frankly, Senator, I am afraid . . . we are going to have very rough sledding."[46]

Thus, the approval and cooperation of the PWA, the Bureau of Reclamation, and Secretary of Interior Ickes were as important as the actual congressional funding. With the help of Wirtz and McDonough, Johnson was able to convince the various governmental departments of the bona fide need to fund additional work in Central Texas. On July 21, 1937, the Department of Interior approved an additional $5 million for funding the Marshall Ford Dam.[47]

The additional $5 million was but seed money needed by the LCRA and the Brown and Root company to accelerate construction at Marshall Ford. Johnson, as congressional representative of the Tenth District, was to take an even more active role after 1937. While work went forward on the first stage of the dam, debate over the proper dam height renewed. The height was as much a logical issue in response to the perceived protection against floods as it was a budgetary concern.[48]

Mansfield, Connally, and Johnson maintained a constant vigil. In large part it was this group which, over the next two years, secured nearly $30 million in federal funds to complete Marshall Ford. In order to maintain harmonious relations with Washington, the LCRA re-invited Secretary Ickes to Texas to dedicate both the Buchanan and the Inks dams. Ickes, who was introduced by Lyndon Johnson, used the occasion to attack the private power companies and extol the benefits of "multipurpose developments" such as the LCRA.[49]

During the remainder of Johnson's first year in office, he was involved with not only the LCRA funding from Washington but also a number of day-to-day disputes among the authority, the city of Austin, the bureau, and the PWA. No detail seemed too trivial to debate.[50]

In spite of the political delays to secure funding, construction proceeded ahead of schedule during early 1938. To complete the dam foundation, 750,000 barrels of low-heat Portland cement at an average price of $1.94 per barrel were needed. Such major contracts to the

Delegations from Texas constantly urged federal approval of additional construction funding. James Roosevelt (center) is flanked by Lyndon Johnson and Alvin Wirtz to urge approval of a $5 million relief grant for completion of Marshall Ford Dam. (*Courtesy Lower Colorado River Authority*)

private sector generated hundreds of jobs. At the dam, a crew of over one thousand worked three forty-hour shifts, pouring approximately eighty thousand cubic yards per day. One year after the January, 1937, onset of construction, the dam was 27 percent complete.[51]

Johnson, in constant contact with Commissioner Page of the bureau, remained concerned over the actual cost of Marshall Ford and the need to provide adequate funding. Although estimated, in late 1936, to be $10 million, the cost of the low dam had jumped to nearly $12 million because of a price increase for the twenty-four floodgates. The first estimate was based on the cost of identical gates purchased for Grand Coulee Dam in November, 1936, yet in 1938 when the equipment was contracted for Marshall Ford, the price had increased over 100 percent, from $32,000 to $66,000 per gate. These costs alone increased the estimate over $1 million.[52]

In addition to being concerned about a steady stream of funds for Marshall Ford, Johnson became anxious about the repayment provisions of the PWA and Reclamation Bureau funding to Texas. By early 1938, fully settled in his new congressional job, Johnson began to make arrangements to ensure that neither the LCRA nor the state would have to reimburse future funding under the reclamation law. Johnson secured a written agreement from Page to add the phrase "exempt as to the Colorado River Project, Texas" at the appropriate point in the funding bill in order to exclude the appropriations for Marshall Ford during the 1939 fiscal year from the repayment requirements. If the requirements had been applicable, there would have been difficulty in completing the structure within the original budget of $10 million and, according to Page, "while there is some benefit to irrigation many miles downstream, the benefit is so indirect that contracts for repayment cannot be obtained." Thus, the primary purpose of Marshall Ford was to be flood control. Page assured Johnson that "the Marshall Ford Dam is unique among those we are building in that it has been [made] exempt from the reimbursable features of the Reclamation Act."[53] By way of justification, Page explained to Johnson that, whereas all other Reclamation Bureau dams were owned by the federal government, the bureau had cooperated with the LCRA only "as an agency for construction . . . which is unlike our position" with respect to other federal reclamation projects, "of which the Colorado River Project, Texas is not one."[54]

Ickes, however, had other plans for Marshall Ford. Shortly after Page had assured Johnson that all work would be done at Marshall Ford and that the LCRA would be excluded from repayment, Ickes, basing his judgment primarily on the rising cost of Marshall Ford, abruptly undercut Page's pledge to Johnson. Ickes, using Page as a scapegoat, advised Johnson that "considering the need for economy in Federal expenditures, Commissioner of Reclamation [Page] has recommended that the larger dam not be undertaken at this time."[55] Ickes justified his reversal by citing Page's recommendation as well as the increased cost, estimated to be over $28 million—or nearly $16 million more than for the low dam already under construction. However, Ickes assured Johnson that if "experience demonstrates the need" the initial low-dam structure could be raised to a higher level at a future date. Furthermore, Ickes concluded, due to the "infrequent large floods," the reservoirs at the Buchanan and Marshall Ford dams, if "drawn down in advance of large flows in the Colorado, Pedernales and Llano Rivers," could protect the region. Assuming that adequate protection was or would be soon in place with the completion of the Marshall Ford Dam, Ickes assured Johnson that information of such "infrequent flows can usually be obtained in time to permit the necessary release of the stored water."[56]

Thus, once again the debate over the low versus the high dam was reopened. Based on the rising concern over the reduced funds, Ickes's pronouncement to Johnson was not a complete surprise. John-

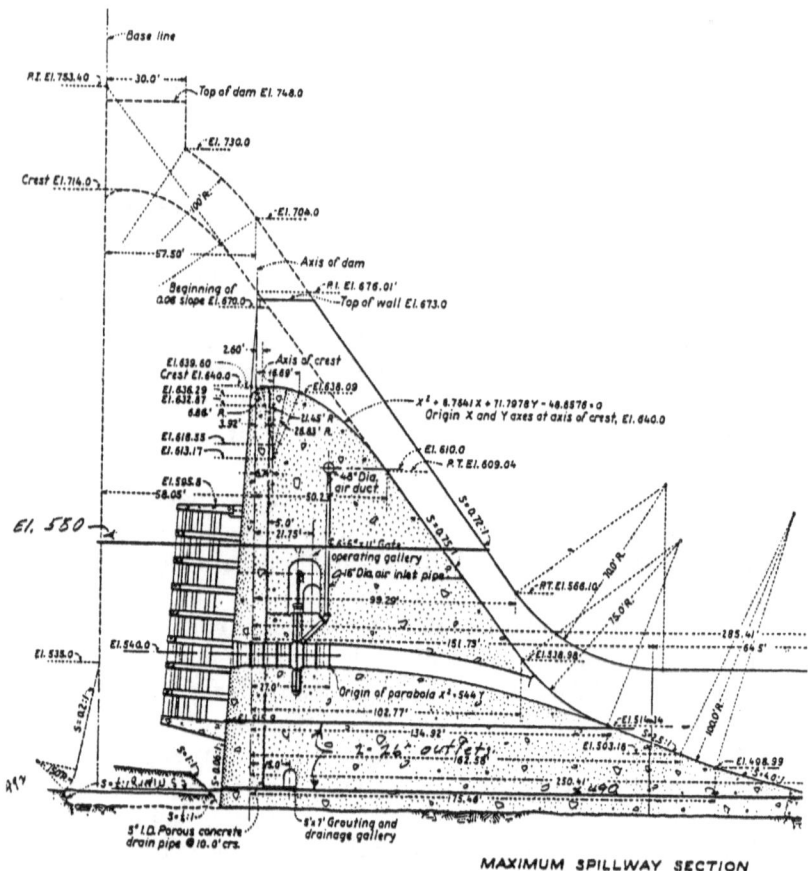

Cross section of the Marshall Ford Dam drafted by the Bureau of Reclamation. (*Courtesy Bureau of Reclamation*)

son and the other members of the Texas delegation analyzed their position and pushed for as complete a project as possible at Marshall Ford. Johnson was satisfied to wait either for a reappraisal of the Marshall Ford request or for the ever-so-frequent "infrequent flow" of the Colorado River to bring attention to the region. Ickes's disclaimer that if "experience demonstrates the need for a higher dam" surely was a prophetic promise to Johnson. The ultimate height of the dam was now in the hands of "mother nature." In due course the erratic nature of the Colorado would play a major role in reinstating Marshall Ford once and for all at its full 265-foot height.

In mid-May John Page announced that Marshall Ford was one-half completed.[57] The work during the spring and early summer was ahead of schedule except for a minor strike and a flood that occurred almost simultaneously on April 27, 1938. Both incidents seemed minor. Floodwaters backed up behind the spillway apron and caused no damage. On the same day the eight cableway signalmen who delivered the trolley buckets of concrete walked out at midnight demanding a raise from $1.00 to $1.25 per hour. The contractor refused their request on the grounds that the minimum wage for cableway signalmen was $.75 per hour. The dispute proved to be ill-timed because the project had to be halted to drain the flooded areas around the construction site. The strike ended at Marshall Ford, but minor labor unrest surfaced among the iron workers and electricians at the Inks and Buchanan damsites. None of these incidents created any extensive work stoppage.[58]

By June total federal funding from the PWA and the bureau for the four LCRA dams exceeded $30 million. Local Austin papers extolled the multipurpose project designed to provide flood control, irrigation, and hydroelectric power—in that order of priority. The loss of over $100 million in property over the past two and one-half decades from the brute force of what the *Austin Statesman* called "Old Man River—the Colorado" was lamented. The multipurpose construction promised to "tame the river considerably."[59]

Refusing to abandon their quest, Johnson and Wirtz, during the first half of 1938, renewed their efforts to raise the Marshall Ford Dam to its full height. Ickes's earlier refusal had not dissuaded Johnson or Wirtz. In June they devised a strategy to promote the high dam—yet allow for compromise on one stage at a time. Alvin Wirtz proved the key tactician. After he consulted with Austin bureau engineer E. A. Moritz, it was determined that the working crew of one thousand could be expanded by a hundred more workers if construction of the high dam was started before completion of the low dam. It was estimated that an additional 440,000 work hours would thus be added to the current estimate of 1.5 million work hours to complete the contract for a low dam. Wirtz and Johnson met with Assistant Interior Secretary E. K. Burlew and Commissioner Page to sell them on a three-part plan. Johnson and Wirtz advocated an allotment of about $15 million to complete the high dam. Failing this, they requested $7 million to bring the high dam foundation to the elevation of the 190-foot low dam under construction. As a last resort, they requested $1.5 million to begin the foundation groundwork for a "future" high dam. The money for the foundation or "first unit," Wirtz assured the Interior Department, would provide immediate employment for one hundred additional men as well as retain the employment of those already at work. Wirtz hoped the $1.5 million could be made immediately available to avoid any layoffs. Once this seed money was received, Johnson believed that Congress could be persuaded at a future date to appropriate sufficient funds for the completion of the high dam.[60]

Page, in late June, advised Ickes of Johnson's and Wirtz's renewed proposals for a high dam. Page was in favor of providing at least $1.3 million for the foundation preparation. He advised Ickes that building the foundation now would simplify the construction of the high dam at a later date; would ensure continued and expanded employment in Central Texas; and save an extra expense of $250,000 if the low structure was completed, including installation of conduits and gates for the present structure that would later be removed to build a foundation for the high dam. Wirtz and Johnson had a strong ally for their case. However, the Interior Department was slow to react to Page's recommendation because the current project already had adequate funding, completion of the low dam was not expected until mid-1939, and congressional action could not be taken for a number of months. Johnson and Wirtz believed that they had achieved victory when Page agreed to draft an application for additional funding. However, they departed for home unsure of the final outcome.[61] What they were unable to do in June, mother nature accomplished in July and August.

In the July, 1938, lull between congressional sessions, Wirtz and Johnson returned to Texas. In an effort to advertise the accomplishments of his first thirteen months in office, Johnson undertook numerous speaking engagements throughout Central Texas. Progress on the Colorado dams was such that Johnson boasted that the fear of flooding was at last eliminated.[62] A few days later, the Central Texas region to which Johnson directed his reassuring comments experienced the worst flooding in over a century. The flooding in late July was the result of over ten days of rainfall that covered nearly the entire forty thousand square miles of the Colorado watershed. The runoff proved more than the Buchanan and Inks dams could safely contain. Without proper gauging or a complete understanding of the hydrology of the upper reaches of the Colorado watershed, the dam operators failed to react quickly enough to the crisis. The Marshall Ford and Austin dams were still under construction and of no help. Irate downstream farmers whose crops and livestock were destroyed and the citizens in Austin were perplexed by their seemingly helpless situation. For example, Austin experienced a high-water level forty-two feet over normal. South Congress Avenue was covered for three days. The resulting devastation and destruction far exceeded that of any previous flood.[63] The multipurpose LCRA dam project designed to protect against floods was instantly labeled a failure.

The months of debate over a high or low dam at Marshall Ford seemed irrelevant to those in flood-ravaged Columbus, Eagle Lake, and Austin. Even before the waters receded, critics were already referring to the catastrophe as the "man-made" Colorado flood. From downtown Austin, Marshall Ford Dam contractor A. J. McKenzie directed sharp jabs at Commissioner Page: "I thought you might be interested in the enclosed clipping—'Flood Damage Angers Farmers!' Even Mr. McDonough seems to be convinced that the high dam at Marshall Ford is necessary."[64] Clearly, the Reclamation Bureau had been cautious in estimating the potential for Colorado River floods and the need for a larger dam. Although Page had reversed his stand a number of times on the issue of a high dam at Marshall Ford, he now quietly, over Ickes's objections, advocated the larger structure. Pressure from Ickes had kept Page from coming out in favor of an additional expenditure. Ickes's earlier admonishment to Johnson that documented "experience" was needed to justify a high dam was now vividly available.

McKenzie and Brown, who maintained that they had been hampered by Washington politics also applied additional pressure for a high dam. McKenzie, eager to gain more funding for Marshall Ford, further admonished Page. As he stood at the window of his Stephen F. Austin Hotel room observing the rising flood on the Colorado he wrote, "Tell us we are not having FUN! Everybody is getting FLOOD CONTROL MINDED, rather than POWER MINDED. You should get some pleasure out of saying, I told you so."[65]

This time the flood on the Colorado was more than what Fritz Engelhard, Chairman of the Lower Colorado River Authority Board, referred to as "one of those things" that happen occasionally.[66] The rising tide of protest stretched all the way from the inundated lower Colorado valley to Washington. Protest meetings were held in numerous cities along the river. In Austin, Senator Connally and Congressmen Johnson and Mansfield heard complaints. Why had not the water been released gradually in advance of the flood? State senator L. J. Sulak, of La Grange, demanded that the state investigate the operation of the dams as he had heard "water was within nine feet of Buchanan Dam spill before it was released." And from inundated Columbus, the local chamber of commerce president, W. G. Dick, asked for and got a reply from Secretary Ickes, who ordered an immediate federal investigation. Dick informed Ickes that property and crop losses in the valley around Columbus exceeded $3 million and that "the people are charging mismanagement to the Colorado River Valley Authority [LCRA]."[67]

There was evidence that the LCRA had possibly been lax in managing water storage at the Buchanan Dam. Most damaging to the credibility of McDonough and the LCRA was a response, published three weeks before the flood, to a concerned public letter. Responding to the claim that Buchanan Lake was already too full, McDonough reported on June 27: "We have worked this solution out and assure you that, with the greatest flood we have ever had above Buchanan Dam, such a flood can be controlled without producing any flood in the lower river."[68] During the flood, farmers in Fayette, Bastrop, Matagorda, Wharton, and Colorado counties recalled McDonough's comments and were irate. Initial reports estimated that there were twelve deaths in the Austin area and over four thousand homeless people. Governor Allred appealed for contributions to help the American Red Cross in their relief work. However, the flood was so extensive and the relief sources so overtaxed that the actual loss in lives and property is still undetermined.

McKenzie continued to badger Page, writing a personal memo stating that the flood "bears out so conclusively the position of the Bureau of Reclamation [to raise Marshall Ford to its maximum height] in regard to flood conditions" in Central Texas.[69] There is no record that Page responded to McKenzie. Quite obviously, McKenzie and the Brown and Root company were eager to continue construction on the high dam at Marshall Ford. To ensure this, Washington had to be convinced of the enormous damage.

The Central Texas flood became headline news nationwide. The interest in the flood reflected the concern of other states over the feasibility of western river projects. The private power utilities were also interested in the investigation of the LCRA, hoping the results would discredit the authority's attempts to produce and sell hydroelectric power. The utilities contended that the LCRA could not be both a flood-control agency and power producer.[70] Price Campbell, president of the West Texas Utilities Company, restated an age-old industry axiom that power production and flood control were conflicting activities "because an empty dam cannot run generators and a full dam cannot store flood waters.... 'You can't catch water in a bucket that is already full.'"[71]

The LCRA was not alone in being attacked by the private utilities. Because the authority received financial aid from the PWA, opponents also attempted to discredit federal participation in the development of public power.[72] Howard Gray, assistant PWA administrator, responding to questions concerning the agency's role in water-development projects concluded:

> I wonder if the propaganda would have been so loud if the dam had been completed and operated by the private utilities as they planned and

The construction of Marshall Ford Dam in 1938, which upon completion was the fourth-largest concrete dam in the world. The Colorado River, shown at a low level in this picture, could quickly rise to flood stage due to upriver runoff. (*Courtesy Lower Colorado River Authority*)

started it.... There have been seven floods on the Colorado River in three years, but this is the first one to be blamed on the PWA instead of the rainfall.[73]

Public sentiment in Texas, once the initial shock of the flood was over, focused on the need for better management of the dam system and the realization that cheap power should not be sacrificed to "power trust propaganda."[74] This was best expressed by a Cisco, Texas, farmer: "This [the LCRA] is the only real opportunity the people have ever had or ever will have to get a square deal on power rates."[75] Cheap and plentiful electricity was becoming a reality, albeit slowly. Lyndon Johnson, in a confidential presentation to the LCRA Board of Directors, had urged prudent but speedy action to identify private industry customers and cities that would purchase the soon-to-be-available LCRA electricity.[76] Yet prior to advocating the sale of power, the authority was subjected to two investigations of the July flood.

As promised, Ickes dispatched one of the most noted engineers of his day, Harry W. Bashore, to review and evaluate the events that resulted in the flooding. Bashore's main charge from Ickes was to determine if the flood was indeed "man-made," a result of substantial negligence on the part of the LCRA. After reviewing the facts, including that the Marshall Ford Dam was only partly completed, Bashore stated that "better preparations should have been made for emergency operations at Buchanan Dam, yet it is doubtful if any careful forethought would have had much effect in reducing the damage covered by the flood."[77] A second internal investigation by the Texas senate was not as brief or as rapidly produced as that of the Bashore report. Begun in mid-August the investigation lasted over four weeks. The state investigation included a full hearing and a cross-examination of LCRA employees, mainly McDonough. Two reports, one by the State Board of Water Engineers, and the second, a *Technical Memorandum on the Colorado River Flood of July, 1938,* by engineer Abraham Streiff, were submitted to the senate panel. The board's report was by far the most detailed and complete. In addition to a thorough review of the characteristics of the watershed and the duration of the storm, the board arrived at twenty-eight conclusions documenting the flood that involved the *entire* contributing area of the Colorado above the Pedernales River.[78]

Less technical in nature yet more political and far-reaching were the interviews and cross-examinations of key witnesses before the select senate panel. Former Gov. Dan Moody presided over the proceedings that mainly pitted Wirtz, representing the LCRA, against the senate investigators.[79] Advocates of the private utilities went directly to the public in an effort to discredit McDonough. They made sure that certain facts were emphasized over and over. The flood was characterized as one of the largest of the century. It was not so much the damage report and technical facts the senators questioned but the intent and philosophy of the LCRA with regard to flood control and hydroelectric power production. After days of hearings the public had a clearer perception of the LCRA in general and the Buchanan Dam in particular. Four primary bits of misleading information surfaced. First, the public was led, by the LCRA and the press during the mid-1930s to believe that the purpose of the Buchanan Dam was primarily flood control. Second, the public was informed (by the authority and in newspapers) that since its completion "there is no need to fear for floods originating above Buchanan." Third, the private utilities put forth a popular belief, contrary to the above two items, that the LCRA "really" intended to use the dams primarily to produce power. Hence, the "flood control talk was merely a subterfuge" to accomplish this purpose.[80] As a result, the attitude emerged that the people below the dam really had very little protection from floods "with the [Buchanan] Dam nearly full."[81]

After hours of hearings, the senate ruled that the LCRA was negligent in its operations of Buchanan Dam and reservoir. The only direct action taken was to refer the committee results to the full senate in order to formulate a clearer policy for use by multipurpose water projects.[82] However, possible full senate action was not nearly so important as the summation presented by Alvin Wirtz. In his closing questions to Alvin Wirtz, Governor Moody formulated an understanding as to the distinct relationship between flood control and power production:

> Governor Moody: Senator, are you in a position to state for the Lower Colorado River Authority with reference to the future operation of these dams so as to give a maximum efficiency in—
>
> Senator Wirtz:—As to the future policy, Governor, the policy—
>
> Governor Moody: Flood control consistent with the production of revenue through the manufacture and sale of hydro electric current in order to meet the maturities on your bond issue?
>
> Senator Wirtz: I think I can state in a general way what it will be, but I can not give you any engineering figure or engineering advice.
>
> Governor Moody: Then I will ask you this question. Is it the policy of the Lower Colorado River Authority in the future to *operate those dams so as to give a maximum efficiency in flood control consistent with the production of revenue through the manufacture and sale of electric current* in order to meet the maturities on the bond issue? [emphasis added]
>
> Senator Wirtz: Why, certainly.
>
> Governor Moody: Then, as I understand it, the future policy of the Board is to operate the dams at a maximum efficiency for flood control consistent with the meeting maturities on your bonds?
>
> Senator Wirtz: Yes. Let me answer this, Governor. I am glad you put the other part in your question, because I think everybody recognizes that the maximum flood control is nothing more or less than that the dam be entirely dry.
>
> Governor Moody: You agree with the policy to that extent.
>
> Senator Wirtz: I think everybody would agree with you on that.
>
> Governor Moody: Then is it not the policy of Lower Colorado River Authority to dump on these people that live down the river, dump and turn out a lot of water in the operation of these dams—

Senator WIRTZ: You do not have to ask me that question. I answered that when I appeared before the mass meeting.

Governor MOODY: I didn't attend that.[83]

Thus, there emerged an understanding that the Lower Colorado River Authority would give first priority to prudent flood control and second priority to generating power to raise revenue in order to retire its outstanding bonds. With the construction phase nearly completed by early 1939, attention turned from building to the potential of electrical distribution throughout Central Texas. It was this mandate, to provide cheap electricity, that truly allowed the Lower Colorado River Authority to stand on its own.

5.
COLORADO LIGHTS:
TEXAS' LITTLE TVA

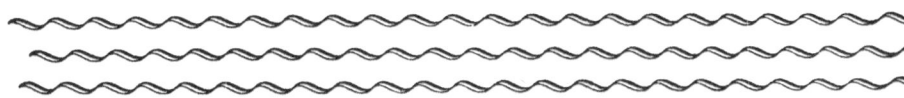

> *The simple disposal of the majority of prime power to private utilities is a crazy idea, I tell you that if we come to such an eventuality we are all likely to be run out of the state and we shall deserve what we received in condemnation for our failure.*
> — Lyndon B. Johnson
> April, 1938

The Colorado River flood of 1938 focused the attention of the state and the nation on Texas and prompted a reevaluation of LCRA projects. In the wake of extensive federal and state investigations, plans were formulated by the LCRA for the distribution of hydroelectric power, initially available from the Buchanan Dam and powerhouse beginning in the fall of 1938. The LCRA, primarily as a result of Lyndon Johnson's stern admonishment, slowly prepared itself to make the transition from the construction phase to marketing electric power throughout Central Texas. Flood control remained an important function of the authority, but producing and distributing power became a coequal function. However, identifying a profitable outlet was an urgent task since the authority needed to raise revenues from the sale of electricity to meet its pending state and federal debt obligations. It was this transition to profitable electric-power production that occupied the authority and its advocates until the eve of World War II.[1]

Ironically, the LCRA realized that its potentially prime case customers would be its enemies — the privately owned utilities companies. While the power companies argued on the one hand that the authority could not be both a flood-control agency and a power producer, on the other they quietly moved to corner the market of *cheap* LCRA electricity when it became available. In reality the question was not whether power would be available but who would control distribution. Yet, flood control and the lessons learned from the flood of 1938 could not be ignored.

These lessons called for regulation at several levels. The Texas legislature had to decide if corrective action was needed to regulate the LCRA and its management of reservoir levels. State and local governmental authorities had to make decisions regarding power-distribution decisions. How was this new power to be distributed, to whom, and at what cost? To have sales required customers with access to electrical power, which created concern over distribution and transmission lines. A coordinated plan developed by Lyndon Johnson, Alvin Wirtz, Max Starcke, the LCRA board, and rural farmers resulted in efforts at both the local and national level to ensure delivery of cheap, plentiful electricity.

The possibility of additional guidelines being passed by the Texas legislature was not simply a reaction to the final 1938 flood report. At stake were millions of dollars of federal funding from Washington especially for completing the Marshall Ford Dam, constructing powerhouses, and purchasing or building transmission lines. Thus, ill-timed action by the Texas legislature in 1939 could affect more than the water level behind Buchanan Dam.[2]

Legislators in Austin recognized the grave importance of their actions. They resolved to settle the issue of the height of Marshall Ford. The flood of 1938 intensified their concern over federal funding in 1939 and 1940. In Washington, Johnson urged cooperation and continued to pressure the Interior Department to accept the LCRA's plans for a high dam. Johnson appealed to Secretary Ickes's political sensibility: "We are never going to have a minute of freedom from the attacks of our enemies as long as we can't do something of consequence to control the Colorado River."[3]

Johnson stressed to federal officials in Washington the need for a completed Marshall Ford, warning that some Texans were advocating draining the Buchanan reservoir in order to have it empty for future flood conditions. He was concerned that the Texas legislature would badly damage the multipurpose-dam concept by taking such quick, shortsighted action.[4]

However, once presented with the realities of the situation, cooler heads prevailed. The senate committee investigating the 1938 flood made no specific policy recommendation when forwarding its report to the full Texas senate. Thus, a bill sponsored by the utilities lobby was introduced in the legislature to keep the reservoir behind the Buchanan Dam *half* empty except at flood periods. The bill died a slow death after receiving an unfavorable report from the House Committee on Criminal Jurisprudence. Committee opposition was based primarily on the contention that the legislature, by imposing limits on the Buchanan Dam reservoir, could damage a good working relationship between the state and the Public Works Administration. Furthermore, opponents of the bill, led by Wirtz and the LCRA board, contended that the Marshall Ford Dam, when raised to its full height, would provide more than adequate flood control.[5]

Johnson's constant political pressure accelerated plans for the high, flood-control dam at Marshall Ford. In late March, 1939, he received the last and final signal of success from Commissioner Page and the Interior Department. The ultimate cost of the dam would be over $30 million; yet to those in flood-prone Central Texas the cost was but a small price to pay for harnessing the river. As further confirmation of his victory in securing needed funding, the April, 1939, issue of *The Reclamation Era* carried a full-cover picture of the construction at Marshall Ford. Quite naturally, Johnson claimed the victory as his very own. Having been in Congress barely twenty-two months, he had made a name for himself not only in the Tenth District in Texas but also in Washington. So impressive were his efforts that he ran unopposed in both the 1938 and the 1940 congressional elections.[6]

The appropriations for the Bureau of Reclamation's construction projects in fiscal year 1940, which began on July 1, 1939, consigned nearly 10 percent of the $65 million budget as an ongoing payment to complete Marshall Ford. Of the key thirty-two projects in twelve western states, only two—the Central Valley project ($23 million) in California and Grand Coulee Dam ($10 million) in Washington State—received more funding. Moreover, Marshall Ford was not the only two-stage (low and high dam) construction. Grand Coulee Dam was similarly built in two stages.[7]

The completion on August 28, 1939, of the "change order" or contract extension to allow Brown and Root, Inc., to continue construction at Marshall Ford pleased the LCRA and all Central Texas. This action along with the ongoing work at the Austin Dam—at Johnson's suggestion renamed the Tom Miller Dam in honor of the mayor—guaranteed continued employment and much-needed money to stimulate the Texas economy. When compared to national statistics, unemployment rates in Texas were generally lower than the national average.[8]

Rural Texans continued to make the best of the poor economic situation in the 1930s. Life in Central Texas around the dams was not any different. Without electricity, ranchers worked from dawn to dusk to complete their chores. Kerosene lanterns and candles provided the only lighting available. Just as significant for rural folks, the lack of electric power meant an absence of comforts and timesaving conveniences. There were no electric appliances, no radios, or—except for those powered by hand or the wind—no pumps to provide household water. Life was not so much primitive as it was spartan.[9] George Norris, concerned about the availability of rural power, commented to Morris Cooke:

> In many ways the farmer is the best customer in the purchase and use of electric current. The farmer ... needs electric current to grind feed, to pump water, and to perform many other services about the farm which have *no* application to the consumer of electricity who dwells in the city.[10]

In order to highlight these circumstances, the LCRA public relations office, beginning in late 1938, released a steady flow of news items aimed at emphasizing the amenities provided by electricity. Comparisons were important to "selling" Central Texans on the need for LCRA power:

> Within two years after Pulaski, Tenn., a city of 3,400 in Giles County, began purchasing power from TVA, the average residential customer had increased his use of electricity by about 150 per cent. Commercial use of electricity rose from 115 kilowatt hours to 262 from October 1934 to October 1936, while the rate dropped from 4.87 to 2.70 cents. Some Texas cities of this size are still paying as much as 10 cents per kilowatt hour.[11]

Thus, LCRA publicists could draw upon the arguments made consistently by public-power advocates such as Morris L. Cooke, George Norris, Judson King, and Gifford Pinchot.[12]

As an advisor to Roosevelt, Morris Cooke echoed a progressive theme which held that abundant electrical power was the property of the people, a resource which would promote a better way of life.[13] Statements by key New Dealers reflected this concept. In an address on electrification of the American home and farm, David Lilienthal, director of the TVA, stressed that "the progress of the country can be traced in our successful efforts to relieve men and women of backbreaking toil and deadening drudgery."[14] Such drudgery could be overcome with widespread rural distribution of electricity to power time-saving machines.

However, it was only in the populated areas of Central Texas, such as Austin, Burnet, Columbus, and Marble Falls, that generator-produced power was available and then only on a limited basis—

usually from dark to midnight. The private utilities claimed that efficiency dictated that they concentrate their efforts almost exclusively in the urban areas. It was their argument that rural areas did not "justify" distribution to a larger rural area. Their reasoning: Why invest new capital in sparsely populated regions, when urban areas provided an easily serviced market at a low capital investment once the generators and lines were complete? A need and a demand in Central Texas were not being met by the private utility companies. Texas political leaders and public-power advocates again looked to Washington for help. Both compromise and cooperation would be needed to distribute electrical power generated by the LCRA.[15]

One of the early objectives of the New Deal was to make electricity more widely available to both rural and urban areas at a fair price. In order to accomplish wide-scale distribution, Roosevelt, by Executive Order no. 7037, established the Rural Electrification Administration (REA) in mid-1935. Initially directed by Morris Cooke, then by John Carmody after 1937, the REA was designed to make self-liquidating loans to communities and local cooperatives organized by the farmers and ranchers.[16] Loans were to be repaid within twenty-five years at 3 percent interest. The only major drawback for Texas, as well as for most other western states, was the stringent population-density requirement to secure an REA loan. Loans would not be provided unless the area served had an average of at least three farms or ranches per mile, and the area in Central Texas had fewer than this required number. In late 1936 a Central Texas delegation petitioned the REA to waive the density rule; but their application was refused, even though, as late as mid-1938, the REA reported that Texas ranked "well down among the states in percentages of electrified farms."[17] In 1935, prior to establishment of the REA, only about 11,000 Texas "farmsteads" had electricity. By 1938, although behind the national average, the spread of rural power had jumped 172 percent, represented by 29,400 rural accounts. The REA concluded that there was a "keen interest in and enthusiasm for" electrical service in rural Texas.[18] Yet, in 1937–1938, as the REA spread in other regions, the situation seemed hopeless for Central Texas. Private utilities were too concerned with the heavy investment and low financial return to expand transmission lines, and the REA was too restrictive to fill the demand.[19]

The REA in Central Texas was not accomplishing what it was created to do, and it was ironic that, despite extensive public power expansion in the West, rural residents, many of whom lived in sight of the dams and power facilities, could not obtain electricity. The trans-Mississippi West was not effectively served by the REA between 1935 and early 1937. Amended REA loan requirements were adopted in 1937 that reduced density constraints and gave some relief, but not on a scale needed in the West. Central Texas would have to wait over two years before the requirements were either changed or waived. In order to secure urgently needed REA funds, Johnson in mid-1938 bypassed Administrator John Carmody and began to appeal directly to President Roosevelt for assistance.[20] Johnson knew that the president had a long record of supporting rural development. It was their deep commitment to natural resource development that engendered harmony between FDR and LBJ.

Roosevelt's dream of promoting economical and plentiful public power dated back to his governorship of New York.[21] His involvement in conservation and multipurpose development of natural resources can be traced to a number of innovative projects. He championed the management of the Saint Lawrence seaway and the creation of the New York Power Authority (NYPA). Power generated by the river would be distributed at low rates by the NYPA throughout rural New York.[22] In order to accomplish expanded public hydropower for the state, Roosevelt appealed to the federal government for cooperation and financial assistance. Although the Hoover administration refused financial help, it did cooperate in order to obtain an agreement among the United States, Canada, and New York State concerning the seaway development.[23] The concept of expedient conservation and large-scale development of cheap, public hydropower was to become a critical theme extending throughout Roosevelt's tenure as governor and into the New Deal of the 1930s.

Roosevelt's early efforts on behalf of multipurpose water development and conservation had wide popular support. To ensure the success of the New York conservation and power plan, he appointed utility expert Morris Cooke to the NYPA board. Cooke boasted after notification of his appointment that it was time to use the Saint Lawrence project as an example to "lift the lid off [the private] utilities."[24] Others, such as John Bauer of the American Public Utilities Bureau, a strong supporter of Roosevelt's and Cooke's efforts, distributed vast amounts of pro–public-ownership literature in order to educate consumers on the benefit of cheap, mass-produced hydroelectric power.[25]

Even while Roosevelt and Cooke were developing a multipurpose river program on the state level in New York, they carefully watched developments in the Tennessee River Valley.[26] It was Roosevelt's commitment to public power that led him as president to support first the TVA and, later, the multipurpose development of streams in other parts of the nation. Although constantly under attack by the Republicans, large private utilities, and coal producers, the TVA slowly emerged as one of the shining stars of Roosevelt's early New Deal.

After his election in 1932, and prior to entering office, he accompanied Senator Norris on a visit to Wilson Dam in central Tennessee.[27] Norris, then past seventy, was both pleased and amazed that Roosevelt, in time, planned to go even further than he had dreamed possible to make the Tennessee River the model public, multipurpose hydroelectric project. Standing in the morning mist overlooking the river and the Wilson Dam, Norris was gratified. For over two decades he had fought against those who wanted to sell the Muscle Shoals project outright as a "war relic" or who wanted to produce power and deliver it to private companies for their distribution.[28]

Roosevelt's intentions and plans mirrored Norris's. If private utilities failed to provide cheap and reliable power, then public authorities, in cooperation with local municipal governments and cooperatives, would produce, market, and distribute electrical power.[29] Roosevelt strongly supported the TVA bill as it was introduced and passed along with other early New Deal measures. From Roosevelt's perspective, the TVA was also intended to promote economic recovery. It would stimulate employment, aid the regional industry, promote navigation, raise the standard of living, and provide cheap electricity to rejuvenate an entire six-state region. In essence the TVA, like the LCRA, marked a unified approach to regional problems by a cooperative effort between federal and local authorities. The president let it be known that he wanted neither restrictions on the authority's capacity to transmit electricity nor any limits on the TVA's building or obtaining its own transmission lines.[30]

Thus, the TVA in the eyes of many in Texas and Washington was an example for the LCRA to follow. The LCRA in matters of electrical distribution copied many of the concepts developed in the early 1930s by the TVA. For example, once created, the TVA was encouraged by Roosevelt to have the widest regional economic impact possible, yet not attempt to duplicate existing power facilities, public or private. More important, the TVA Act mandated that a "preference" be given to public distribution agencies such as rural cooperatives. The preference concept dated from the early 1900s, in the years before World War I during the earliest debate between public and private power interests over the Hetch Hetchy project east of San Francisco. There had been numerous interpretations of the concept over the years, and they continued to develop long after the coming of the New Deal. The purpose of the "preference" principle was to allow public electric power organizations to develop in opposition to private utilities. Consequently, the TVA became the "yardstick" for nationwide electrical rates, measuring not only power rates but also the degree to which the multipurpose concept could be applied to a unified system. Thus, the TVA had an impact on all multipurpose river projects that followed.[31]

Like most budding public power producers of this era, the LCRA, referred to by many supporters as "Texas' little TVA," was keenly interested in developments in both Washington, D.C., and along the Tennessee River. TVA director David Lilienthal attempted to minimize friction with the regional utilities, even though a decade-long struggle ensued over jurisdiction, distribution, and rates. The low rates offered by the TVA were its primary weapon in facilitating the spread of low-cost power.[32]

Slowly, private power companies in the Tennessee valley realized that cooperation was the best way to ensure their own survival. In Central Texas the case was not unlike that in the Tennessee valley. The multipurpose approach had provided flood protection with the completion of the high dam, Marshall Ford, which was renamed Mansfield Dam in honor of Congressman Joseph J. Mansfield. Now what was needed was a demand for the new power being generated, first, at Buchanan Dam and, later, by other LCRA dams. Lyndon Johnson liked the TVA idea of the "yardstick-for-power" philosophy. However, not everyone in Texas agreed. Mayor Tom Miller of Austin was described as strongly opposed to the lower rates for other reasons. "[Miller] was still looking at the prospect of what's going on still today, the transfers of large blocks of money to keep the tax rate low, spread it across. Tom always used the argument, which is still a valid argument, that all the tax-exempt properties in Austin, that they use electricity, so we'll tax them through the electric system." Following the president's lead, Johnson protested against the initial proposal by the LCRA to sell direct to the private utilities, calling it "a crazy idea," even though they pledged to buy all the power the authority could produce.[33] He made his views widely known throughout central Texas:

> I have fought to hold this vast treasure of cheap electric power for the people until such a time as they could decide how they wanted to take it. I have done everything I could to see that all the facts were brought to their notice. All the time I have been doing this, and others as deeply interested as I have been on the job, the power companies have been telling the people they didn't know how to run their own business and ought to let New Yorkers continue to drain Texas of its resources.
>
> I believe that the Colorado River belongs to all the people. Nature gave it to them, to be a blessing as well as a curse, if they had the ingenuity to convert a curse into a blessing. They did have. The water in the river belongs to the people. All the LCRA dams belong to the people. And the electricity which is generated will also be the people's.

Now that we have brought about . . . an ample supply of cheap power to build Central Texas into a strong and prosperous section of the state, we are asked to turn that power over to private companies, controlled entirely by New York bankers, so they can sell our power back to us at 200 or 300 per cent profit, or at any price they think the traffic will tolerate.

Last year alone Central Texas paid private utility companies owned and controlled on Wall Street, more than $1,000,000 in excess charges. That million came out of the pocket of every person in the Tenth Congressional District who had so much as one electric light socket in his house. If Central Texans had been using their own cheap hydroelectric power last year, instead of the expensive private-utility power at famine prices, they could have had just as much electricity as they used and put $1,000,000 in the bank for a rainy day, besides.

Water can run through generators as well as it can get through gates, and just as fast. One way it has to earn its pay, the other it goes off gallivanting like an untamed horse. By means of our conservation program we have a fine supply of electric power. That power has been created. It is going to be sold. It will be sold either directly to the people who own it, whose water generates it, at a cheap cost price, or it will be sold at a cheap price to the power trust which will dish it out like jewelry and diamonds to the people who own it, and charge them all they can squeeze out of them.[34]

At the request of Alvin Wirtz, Johnson toned down his public opposition to private utilities and cooperated with the LCRA in working out an agreement with the private utilities in Central Texas. Due primarily to the lack of an extensive network of transmission lines, it proved to be in the best interest of the LCRA to avoid a "fight" with the private companies and buy outright the existing lines and stations. Wirtz assured Johnson, "I think the best thing that could happen to us would be for PWA to give us the allotment for a transmission system. . . . The battle is getting a little hot but there is no reason to think we will not win."[35] Wirtz emphasized that purchase was preferable to duplication of the transmission systems. Johnson, by appealing directly to the president, helped speed approval of the needed PWA funding.

In late 1938 Roosevelt approved a PWA allotment for electric distribution in the amount of $459,922.[36] In September, 1938, the Pedernales Electric Cooperative received a loan of $2 million from the REA to build 1,830 miles of transmission lines in the Hill Country. In order to put these funds to use there could be no duplication of lines. The ranchers and urban dwellers to be served were to assist

Lyndon Johnson in early 1938 with members of the first Pedernales Electric Cooperative Board. Johnson was a strong advocate of cheap, plentiful rural electricity. (*Courtesy Lower Colorado River Authority*)

with bond elections to guarantee repayment. The PWA allotment spurred community involvement. However, Johnson and LCRA officials were at first frustrated by rural ignorance. Never having had electricity, many were afraid of the technology. They were intimidated by the seeming complexities of creating a co-op. Johnson went from city to city and from county to county to promote the concept and its benefits.[37]

In most cities and rural areas the choice was put to a vote. This process created some friction, for it pitted local private interests against what seemed to be the "distant" and "far-off" LCRA network. To improve public relations, Johnson encouraged the LCRA to create a newspaper, *Colorado Lights*, to provide comparative information on rates, distribution, and the benefits of "electrification."[38] In addition to the travel and newspaper promotions, Johnson went on the radio, on station WOAI in San Antonio, to win support for inexpensive public electric power:

> There are two ways the LCRA can sell its power, the first way is to sell their power—the people's power—outright to the private utility companies which will turn right around and sell it back to them at a profit.
>
> The second way is to sell the people's power to the people themselves at cost. They can buy this power at about one-half cent a kilowatt hour. For a maximum of about 3½ cents a kilowatt-hour, they can distribute this power, pay all operating costs, pay off their indebtedness and interest, and have a profit, besides, to help ease their tax burdens and reduce their rates.³⁹

Sensing that they faced defeat, the private utilities in Central Texas began in 1939 to reevaluate their position in light of the LCRA's success in obtaining PWA funds. Realizing that such strong public-power advocates as Johnson, Wirtz, and Starcke would just as soon duplicate transmission lines, the private power companies, led by the Texas Power and Light Company, reached an agreement with the LCRA. In one bold move, using REA and PWA funds, the LCRA bought from the TP&L the transmission lines and equipment in a sixteen-county area in Central Texas. City after city agreed to "tie on" to LCRA power.⁴⁰

There were overwhelming incentives to do so. LCRA rates were 37 percent below those of private companies.⁴¹ In a somewhat conciliatory scheme the LCRA sold excess power in such distant markets as Houston (a market not originally included in the service area) via the Houston Lighting and Power Company at the same rate as that given to towns in "public ownership" areas. In order to protect its future electricity needs the LCRA stipulated a "recapture clause" in its ten-year contract with the privately owned utilities. Thus, extra power could be produced and sold outside the LCRA service area unless demand warranted only local sales. The success of the LCRA caught the attention of many, including the president.

Pleased with the outcome of the LCRA's ability to secure the TP&L properties, Roosevelt wrote to Lyndon Johnson in late July: "It shows it is possible for a neighboring private utility to cooperate with a public power development to the advantage of both and the public."⁴² More important, Roosevelt cooperated with Johnson in persuading the REA to advance a loan to Central Texas cooperatives.

Shortly after Johnson's success in aiding the LCRA, Roosevelt offered him an appointment as administrator of the Rural Electrification Administration. Although unanimously recommended to the president "from all sources," Johnson declined the appointment, citing unfinished business in the Tenth District as his reason. His refusal set in motion a curious turn of events.[43]

In September, 1939, Undersecretary of Interior Harry Slattery accepted the post of REA administrator. His departure from Ickes's staff left open a position for which the secretary recommended Alvin J. Wirtz of Texas.[44] After Wirtz met with Secretary Ickes and briefly for an interview with the president, Ickes recorded his approval by commenting in his diary: "His handling of the Lower Colorado River project [LCRA] has been masterly, both as a lawyer and as a negotiator."[45]

Ickes made good use of Wirtz's knowledge as an authority on flood control, irrigation, reclamation, and water power by having him undertake a speaking tour of the West in order to reinforce the Interior Department's plans. For over a year Undersecretary Wirtz used the experience and knowledge gained in Central Texas to help oversee the completion of the Grand Coulee and Shasta projects, the two largest hydroelectric dam projects in the world in 1940. An able representative of the administration, he noted in Spokane, Washington: "It is logical that states and communities cooperate with the Federal Government in the development of large projects. Such a partnership will serve the common interest and goals throughout the West."[46] In a later speech in Portland he used the Texas example to promote federal involvement at the local level:

> Now, down in Texas for many years we prided ourselves on the fact that we were independent, that we didn't want the Federal Government to have anything to do with us down there, except maybe once in a while to build a post office.
>
> We believed in the principle of absolute states rights until we woke up and found that all we had was the principle. Then we began to look and see if we could find the financial means and resources of rehabilitating ourselves.[47]

After a hectic tenure as undersecretary in the Interior Department and Texas Democratic Party coordinator for Roosevelt's 1940 reelection campaign, Wirtz resigned in mid-1941. In a letter to George Norris, he revealed his reason: he wished "to participate in the campaign [for the Senate] for our young friend, Lyndon Johnson."[48] Wirtz, at the height of his political power and influence, advised Norris in a personal letter from Austin: "Although it has been a hard uphill fight, we have our 'Little T.V.A.' functioning along the pattern set by its big brother."[49] Part of Wirtz's strategy for Johnson's 1941 bid for a Senate seat was to point out the good Johnson had done for Central Texas in securing funding for the LCRA and rural electricity. However, Johnson, running with full White House backing and on the slogan of

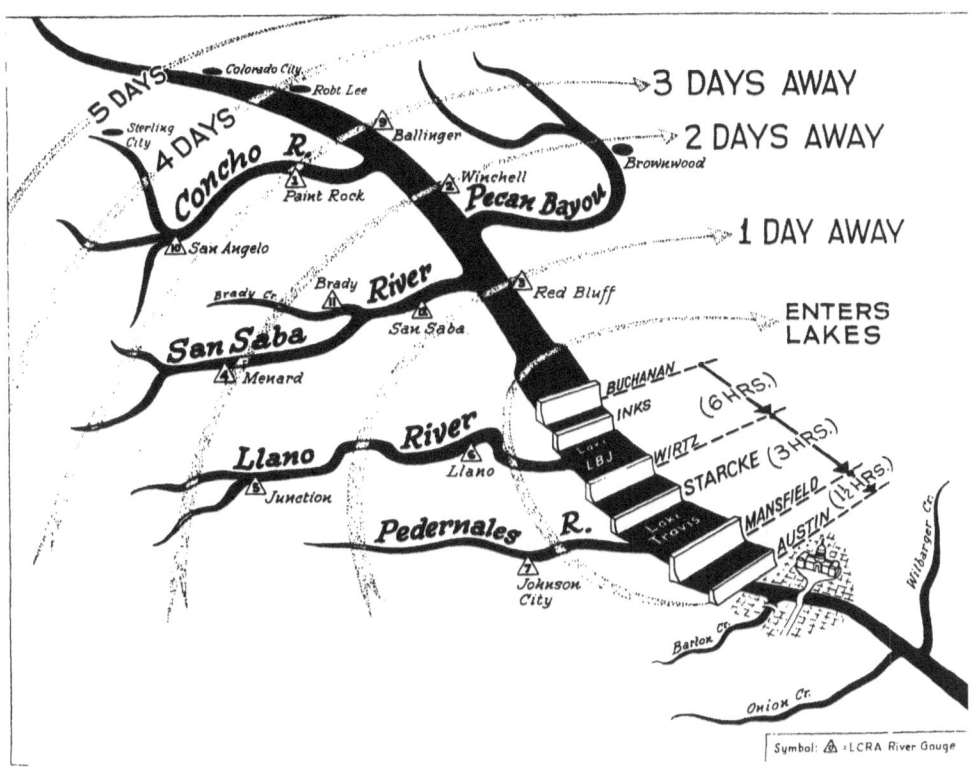

Major dams and the watershed of the Colorado River above Austin. (Courtesy Lower Colorado River Authority)

"Roosevelt and unity," lost to Gov. W. Lee ("Pappy") O'Daniel in a heated election decided by 1,311 votes out of nearly 600,000 cast. Gravely disappointed, Johnson returned to his House seat, where he continued to be an avid supporter of the LCRA.[50]

In addition to Wirtz and Johnson, others connected with the LCRA project also helped disseminate the lessons learned in Central Texas to other regional hydroelectric projects. The LCRA hosted numerous visitors, including a delegation led by David Lilienthal of the TVA.[51] Engineers from the LCRA project took jobs at Hoover, Parker, Shasta, Bonneville, and Grand Coulee dams.[52]

The initial four dams—Buchanan (Hamilton), Mansfield (Marshall Ford), Tom Miller, and Roy Inks—had a combined hydroelectric power output of 127,500 kilowatts in 1941.[53] Two additional dams—Granite Shoals (Alvin Wirtz Dam) and Marble Falls (Max Starcke Dam)—were planned in 1939 but not added to the LCRA system un-

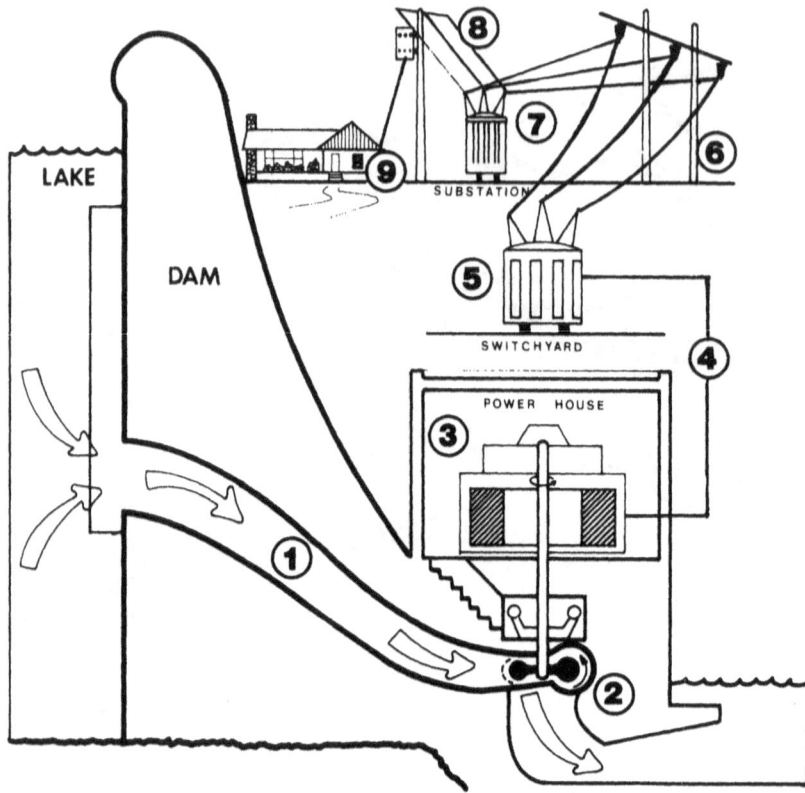

Hydroelectric dams produce power when reservoir water flows down a large pipe called a penstock (1) to the turbine (2), where it turns the impeller blades that are fixed to the turbine shaft inside the generator (3), which converts the motion of the shaft into electric energy. The water flows back to the river, while the electric energy flows through metal conductors called buses (4) to a switchyard (5). There, transformers "step up" the voltage, causing electricity to move out over miles of power lines (6) to a substation (7), where another transformer "steps down" voltage for passage over smaller lines to an electric pole transformer (8), which reduces voltage for home use (9).

til after the war. The electrical power produced by the powerhouses adjacent to the dams served an area of 31,000 square miles and transformed the rural Central Texas lifestyle. In addition to the creation of the LCRA, the PWA and REA deserve equal credit for the rapid spread of rural electricity. Texas, which ranked first in the number (over 500,000) of farms and ranches in 1934, had rural electric service to only 2.3 percent of these "farmsteads." Elsewhere in the nation

TABLE 2. **Lower Colorado River Authority Facts, 1941**

	Dams			
	Buchanan (Hamilton)	Roy Inks	Mansfield (Marshall Ford)	Austin (Tom Miller)
Height (ft)	145.5	97.0	265	100
Length (ft)	11,000	1,545	4,600	1,500
Dam thickness				
Base (ft)	215.11	76.1	213	115
Top (ft)	34	16.5	30	22.8
Lake size (ac)	23,000	900	27,000	3,000
Water impounded				
(billions of gals)	359.7	4.3	625	7
Water capacity (ac-ft)	1 million	17,500	1.2 million	21,000
Max lake depth (ft)	132	60	225	30
Max lake length (mi)	32	2.8	51	20
Max width (mi)	8	.57	8.5	.5
Dam elev above				
sea level (ft)	1025.5	922	750	493
Miles from				
mouth of river	413	410	318	297
Lake shoreline (mi)	192.0	40	270	100
Power generation				
(kva)	25,000	12,500	75,000	15,000

SOURCE: Lower Colorado River Authority, *What Is the Lower Colorado River Authority?* Austin, circa 1941.

only Mississippi, Louisiana, and Arkansas had a lower percentage of rural service. But, because of the speed with which the REA moved, Texas boasted an increase in "electrified farms" of 516.3 percent by mid-1940.[54] Inexpensive rural power and effective flood control had at last come to Central Texas by the eve of World War II.

As a result of over a decade of forceful political leadership in both Austin and Washington, Central Texas advocates of flood control and plentiful hydroelectric power were vindicated. Behind the dreams and efforts of Buchanan, Mansfield, Rayburn, Wirtz, Miller, and Johnson, a consensus existed to overcome the delays and conflicts posed by opposition both in Texas and Washington, D.C. Capitalizing on federal financial programs and assistance, the Lower Colorado River Authority—"Texas' Little T.V.A."—fit neatly into the New Deal's overall scheme of advancing the multipurpose river development.

Until late 1942, when the lignite-powered Comal Plant was acquired, the hydroelectric power dams supplies 100 percent of all electricity generated by the authority. As demand grew, newer and more efficient systems were added to meet the demands of central Texas.[55] Yet, as the LCRA shifted from a limited to a greater emphasis on hydroelectric production, advocates were quick to point to the early cooperation between Texas and the federal government in developing the multipurpose network of dams. In 1957, on the eve of the twentieth anniversary of Buchanan Dam, Sen. Lyndon Johnson noted:

> A dam is more than a collection of concrete and steel, laid out with plumb bob and transit and constructed according to sound engineering principles. It is also a human achievement of the highest order.
>
> It is a vindication of the struggle of the past; a source of comfort and prosperity in the present; and a pledge of progress and growth for the future.[56]

6.
CONCLUSION

The creation and development of the Lower Colorado River Authority proved to be a timely boost to the economy and growth of Central Texas. The impact of New Deal programs supporting reclamation, conservation, and power-development projects, totaling over $2 billion in the trans-Mississippi region, aided the West (including Texas) in recovering from economic stagnation and depression. Based largely on an effort to manage and develop natural resources and inland waterways more effectively, New Deal programs turned many historically flood-prone rivers into multipurpose enterprises that provided flood control, a readily available water supply, and electrical power. The major thrust of the New Deal water projects in Texas was one of economic rejuvenation facilitated by federal-state cooperation.

Due in part to the magnitude of multipurpose river projects, state and private efforts alone could not have alleviated large-scale flooding and promoted local economic recovery. During the 1920s and early 1930s private industry was given ample opportunity to develop dams, yet could neither muster the resources needed to conquer the river nor meet the capital investment demands of such large projects.

Although the intention was to have private companies and/or state agencies develop the Colorado, the Great Depression made this plan unattainable. Texas was forced to come to grips with the fact that federal support and cooperation could, at least in the short term, improve the declining economic conditions in Central Texas. Federal-state cooperation, thus, became a viable option.

Texas was fortunate in having a seasoned group of legislators in Washington who had no fear of federal intervention in state and local

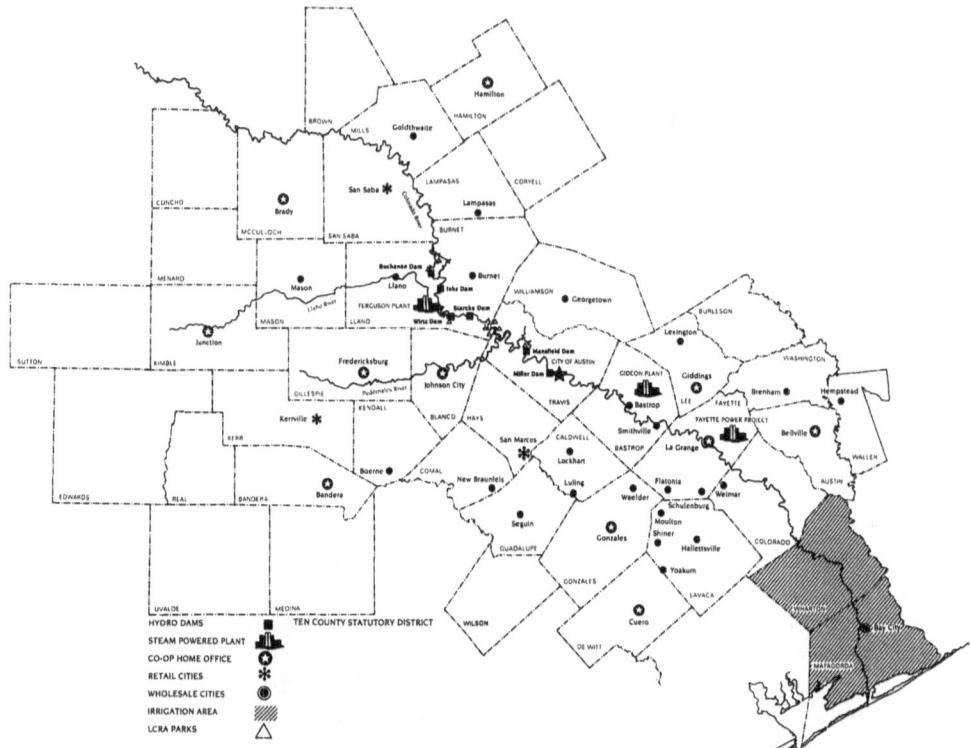

The statutory area of the Lower Colorado River Authority covers ten counties: San Saba, Llano, Burnet, Blanco, Travis, Bastrop, Fayette, Colorado, Wharton, and Matagorda. *(Courtesy Lower Colorado River Authority)*

affairs. This key corps of leaders included James P. Buchanan, J. J. Mansfield, Sam Rayburn, and young Lyndon B. Johnson. In addition to their Washington expertise, they retained the political and personal connections in Texas needed to facilitate the timely implementation of federal New Deal programs. The Democratic leadership from Texas (including Vice President John Nance Garner, who broke ranks with Franklin D. Roosevelt in 1937) was well positioned to influence and foster the growth of the Lower Colorado River Authority and other New Deal activities in Texas.

Although Texas was well represented at the federal level, local politicians more often than not found themselves in conflict with broader federal goals. Frequently intent upon preserving the status quo, they perceived New Deal programs as transferring power to Washington. Consequently, a coalition of diverse interests represented by Austin Mayor Tom Miller, legislator Sarah Hughes, and the West Texas ranchers' lobby organized to combat the LCRA. However, the rewards promised and delivered by the LCRA were so far-reaching that the opposition was overwhelmed. The LCRA is thus an excellent case study of federal-state cooperation that spread throughout the West in the 1930s.

In the development of the LCRA, such cooperation was both expedient and rewarding to the economic welfare of Central Texas. The promise of jobs and economic stimulation went hand in hand with the earlier promise of flood control and eventually cheap hydroelectric power. The impact of the New Deal on the West, as well as Central Texas, was much more profound than the mere economic dimension. States of the trans-Mississippi West covered under the Reclamation Act of 1902 were able, with the help of New Deal programs directed by the Bureau of Reclamation, the Public Works Administration, and the Work Projects Administration, to engage in multipurpose river and resource development never before attempted on such a large scale by either private or public means.

Within less than a decade a dozen of the world's largest multipurpose dams were built in the American West. Resource management shifted from a provincial local option in the hands of a few to a centralized regional approach under federal auspices. The LCRA and Central Texas, which for a brief period in 1937–38 boasted the fourth largest dam in the world—Marshall Ford Dam—was but slightly different in that the Colorado River of Texas is totally an intrastate system. Once local political opposition was overcome, the LCRA multipurpose network of dams north of Austin fit neatly into the larger national recovery concept of the New Deal. Federal funds for construction of dams created much-needed jobs and stimulated the economy of Central Texas.

During the early New Deal era, the LCRA represented one of the largest multidam and hydroelectric projects west of the Mississippi River and, on a national scale, second only to the multistate Tennessee valley development. In many respects Texas' efforts on the Colorado River in the 1930s endeavored to do at the state level with federal cooperation (with less friction and conflict) what the Tennessee River project had attempted to do in over two decades of turmoil. Hence, the Tennessee Valley Authority, established in 1933, was to become a model for all such multipurpose projects worldwide. Texans, while taking pride in their own accomplishments, were quick to refer to the LCRA project as "Texas' Little TVA."

On a regional scale the LCRA was a successful project in aiding economic recovery. Furthermore, it and the Reclamation Bureau provided training and on-the-job experience for engineers, draftsmen, and technicians. These skills proved useful at other projects throughout the West. Alvin Wirtz and Lyndon Johnson, however, provided the greatest link between the LCRA and other developments in the nation. Wirtz, the noted water jurist of Texas, became undersecretary of interior in the late 1930s and helped galvanize support for other western water projects. Additionally, Wirtz was a key regional leader in securing Roosevelt's third nomination in 1940. After the election Wirtz left the administration, returning to Texas to promote the affairs of the LCRA and to advise Texas representative Lyndon Johnson on his first bid for the U.S. Senate.

Although defeated, Johnson persevered and secured a Senate seat in 1948. Ever mindful of his early roots in the Tenth Congressional District in Texas—the location of the LCRA—Johnson, throughout his political career, was an advocate of water development, the advancement of rural electricity, and the general improvement of social conditions that national prosperity should foster. In line with Roosevelt, his mentor, Johnson felt that the federal government should do what the states were either unable or unwilling to implement.

And the New Deal in the West *did* foster prosperity. Prior to 1933, water projects, in the hands of local concerns, had limited use and scope. New Deal programs stressed better planning and multipurpose projects, ensured large-scale employment, attracted new industry to the West, and stimulated extensive agriculture. The LCRA is a good case in point. In Central Texas the LCRA network of electrical power distribution, with help from the Rural Electrification Administration via "cooperatives," is a prime example of how rural areas were able to enhance their otherwise spartan way of life. Electricity transformed the West. Economic electrical power was being distributed widely by the late 1930s and early 1940s. Power quickly became the keystone and stimulant for sustaining regional economic recovery long after the construction phase was complete. Hence, today in the West the federal government continues to play a large role in the development and management of our water resources.

The early leadership of the LCRA fully grasped the magnitude of the authority's impact on the region. Often far ahead of their contemporaries, both public and private, administrators of the LCRA, which generated 100 percent of its power from dams as late as 1941, had as early as 1937 anticipated the day when new generating sources would be needed to keep pace with regional growth. Such early planning has enabled the agency to avoid being totally dependent on hydroelectric power and take a multifaceted approach to electric generation. The planning and organization of Tom McDonough and Max

CONCLUSION 113

Even as president, Lyndon Johnson would boast that his contribution to the development of the Lower Colorado River Authority was his "proudest accomplishment." (*Courtesy LBJ Library*)

Starcke (assisted by Alvin Wirtz) is an example of such prudent judgment. An emphasis on water resource "planning" by the LCRA is also reflected in other western water projects and is a true legacy of the New Deal.

Although electricity is hailed as the chief by-product of western river projects, including the LCRA, it is the management and conservation of water resources that will determine the full impact of the "planning ethic" generated in the late 1930s and early 1940s. Today, only 5 percent of the electricity generated by the LCRA is produced by the Colorado River dams. Since 1942 the authority has used coal, gas, and lignite resources to ensure ample power for Central Texas. Given the fundamental purpose for its creation, the authority today continues to serve the larger function of flood control and provider of ample water.

Faced with a series of conflicts, the state and federal governments were able through the LCRA to establish a mutual, cooperative effort to harness the Colorado River and stimulate the economy of Central Texas. Water-resource management, begun on a regional basis with such programs, has profited enormously from the experience of the Lower Colorado River Authority.

Notes

CHAPTER 1

1. *Texas Almanac, 1980–1981,* 92. No two sources agree on the size of the drainage area. There is considerable disagreement on whether certain areas are contributory. Most estimates range from 27,800 to 42,500 square miles. The drainage area of the Tennessee River basin at its entrance to the Ohio River is 40,600 square miles, and the watershed of the Brazos River is 44,640 square miles. See also H. K. Barrows, *Floods: Their Hydrology and Control,* 280–81; C. L. Dowell and S. D. Breeding, *Dams and Reservoirs in Texas;* Kenneth E. Hendrickson, *The Waters of the Brazos: History of the Brazos River Authority, 1929–1979;* Federal Security Agency, *Report on Water Pollution: Colorado River Sub-Basin of Western Gulf Drainage Basin* (Fort Worth, 1952), National Archives Center, Fort Worth, Texas, in National Archives, Record Group 138, Records of the Federal Power Commission, Regional Engineer, Fort Worth, General Subject Files 1930–1954, box 20; H. E. Robbins, Regional Director, to B. L. Robinson, District Engineer, May 19, 1948, Flood Control Regulations, Marshall Ford Reservoir, Colorado River, Texas, in National Archives, Record Group 77, Records of the Corps of Engineers, Fort Worth District, Construction Division—Civil Works Project 1934–1961, box 56, File 821.21, Marshall Ford Reservoir; R. R. Neyland, "A Few Facts about the Tennessee River and the Wilson Dam," found in Tennessee Valley Authority Papers, box 1.

2. U.S. Congress, House of Representatives, *The Water Powers of Texas,* by T. U. Taylor, 58th Cong. 2d sess., H. Doc. 759, 41–43; idem, *Preliminary Examination of Colorado River, Texas, with a View to Devising Plans for Flood Protection,* 66th Cong., 1st sess., H. Doc. 304, 1–12; U.S. Department of Agriculture, *Middle Colorado River Watershed, Texas,* 78th Cong., 1st sess., H. Doc. 270, 1–6. See also Walter Prescott Webb, *More Water for Texas: The Problem and the Plan,* 3–9.

3. *Texas Almanac, 1980–1981,* 272, 339, 343, 372; E. W. Cole, "LaSalle in Texas," *Southwestern Historical Quarterly* 49 (Apr., 1946): 484–85; Henry R. Stiles, ed., *Joutel's Journal of LaSalle's Last Voyage, 1684–1687,* 122–232; U.S. Congress, House of Representatives, *Preliminary Examination of Colorado*

River, Texas, from Its Mouth As Far As Is Practicable, with a View to Removing the Raft, 66th Cong., 2d sess., 1919, H. Doc. 529, 1–23.

4. Eugene C. Barker, "Description of Texas by Stephen F. Austin," *Southwestern Historical Quarterly* 29 (Oct., 1924): 98–121. See also Eugene C. Barker, *The Life of Stephen F. Austin: Founder of Texas 1793–1836*, 23–42, 80–118.

5. U.S. Congress, House of Representatives, *Water Powers of Texas*, 41–43; Stanley C. Banks, "The Mormon Migration into Texas," *Southwestern Historical Quarterly* 49 (Oct., 1945): 236.

6. Ernest W. Winkler, "The Permanent Location of the Seat of Government," *Texas State Historical Association Quarterly* 10 (Jan., 1907): 217–18.

7. Ibid., 207–25; O. M. Roberts, "The Capitals of Texas," *Texas State Historical Association Quarterly* 2 (Oct., 1898): 117–19; C. W. Raines, "Enduring Laws of the Republic of Texas," *Texas State Historical Association Quarterly* 2 (Oct., 1898): 155–58; A. K. Christian, "Mirabeau Buonaparte Lamar," *Southwestern Historical Quarterly* 23 (Apr., 1920): 264–70.

8. Frank Brown, *Annals of Travis County and of the City of Austin: From the Earliest Times to the Close of 1875*, 5; Nan T. Thompson, "The Muddy Brazos in Early Texas,"*Southwestern Historical Review* 63 (Oct., 1959): 239. Early settlers and explorers often confused the Colorado with the Brazos. See also Alex W. Terrell, "The City of Austin from 1839 to 1865," *Texas State Historical Association Quarterly* 14 (Oct., 1910): 113–28.

9. R. L. Lowry, *Flood Control by Marshall Ford Reservoir—Colorado River Project, Texas*, 1–15, 198; *Austin Tri-Weekly Texas State Gazette*, July 12, 14, 1869. Settlers along the Colorado experienced major floods in 1843, 1852, 1869, 1870, 1898, 1899, and 1900. See also Walter E. Long, *Flood to Faucet*, 1–9.

10. Lowry, *Flood Control by Marshall Ford Reservoir*, U.S. Congress, Senate, *Annual Report of the Secretary of War*, 33d Cong., 1st sess., S. Doc. 1, vol. 2, 223, 567; idem, *Annual Report of the Secretary of War*, 33d Cong., 2d sess., Exec. Doc. 1, vol. 2, 167; *United States Statutes at Large*, vol. 01, 57. See also U.S. Department of Army, Chief of Engineers, *Preliminary Examination and Survey of Colorado River, Texas, with a View to Its Improvements by Means of Locks and Dams*, 43d Cong., 2d sess., 1875, H. Doc. 657, 1–8.

11. Charles J. Allen to Thomas L. Casey, Dec. 20, 1890, in U.S. Congress, House of Representatives, *Report of the Examination of the Colorado River, Texas*, 51st Cong., 2d sess., 1891, H. Doc. 138, 2.

12. H. P. Bunger, *Irrigation and Flood Protection, Austin to Matagorda, Texas*, no. 20, 5; State of Texas, *General Laws*, 38th Leg., 2d C.S., 103–105. The act of 1923 highlighted the fact that in the years 1913, 1914, 1919, and 1922 the Colorado floods were influenced by the raft. Each year both Bay City and Wharton were completely inundated. For a detailed presentation see Comer Clay, "The Colorado River Raft," *Southwestern Historical Quarterly* 52 (Apr., 1949): 410–26; U.S. Department of Army, Chief of Engineers, *Preliminary Examination and Survey of Colorado River, Texas*, 60th Cong., 2d sess., 1900, H. Doc. 1211, 1–14; idem, *Preliminary Examination Colorado River, Texas, Having in View the Extent of Any Improvement of Said River*

with *Reference to Snagging and Cleaning the River*, 63d Cong., 1st sess., 1913, H. Doc. 3, 1–9.

13. Long, *Flood to Faucet*, 1–16. The best review on the early development and "raft" is found in Comer Clay, "The Lower Colorado River Authority: A Study in Politics and Public Administration" (Ph.D. diss., University of Texas, 1948).

14. *Austin Statesman*, June 18, 23, 25, 1895; U.S. Geological Survey, *The Austin Dam*, 1–52; U.S. Congress, House of Representatives, *Water Powers of Texas*, 43.

15. *Austin Semi-Weekly Statesman*, Apr. 11, 1900. Prior to the dam failure, the water crested at 11.07 feet over the top of the dam and had a discharge of 151,000 second-feet. See also Long, *Flood to Faucet*, 11–27.

16. U.S. Geological Survey, *Austin Dam*, 47–49; *Austin Statesman*, Apr. 8, 9, 10, 1900; U.S. Congress, House of Representatives, *Water Powers of Texas*, 43–47; idem, *Report of Preliminary Examination of Colorado River, Texas*, 71st Cong., 2d sess., H. Doc. 361, 15; U.S. Geological Survey, Water Supply and Irrigation Papers, no. 28, 118–19; B. F. Thomas and D. A. Watt, *The Improvement of Rivers: A Treatise on the Methods Employed for Improving Streams for Open Navigation, and for Navigation by Means of Locks and Dams*, 307, 505, 671. See *Engineering News-Record* 43 (Feb., 1900).

17. U.S. Congress, House of Representatives, *Report of Preliminary Examination of Colorado River*, 71st Cong. 2d sess., H. Doc. 361, 15; Motl v. Boyd, 286 S.W. 458 (1926).

18. U.S. Congress, House of Representatives, *Report of the Preliminary Examination of the Trinity River, Texas*, 51st Cong., 2d sess., H. Doc. 275, 1–4; idem, *Report of Examination of Trinity River, Texas*, 58th Cong., 2d sess., H. Doc. 118, 1–5; idem, *Report of Examination of Guadalupe River, Texas*, 58th Cong., 2d sess., H. Doc. 187, 1–4; idem, *Report of Examination of Sulphur River, Texas*, 58th Cong., 2d sess., H. Doc. 231, 1–5; idem, *Report of Examination of Sabine River from Its Mouth to Logan's Port, LA, and Brazoria, Texas*, 60th Cong., 1st sess., H. Doc. 490, 1–4; idem, *Report of Examination of Neches River, Texas*, 60th Cong., 1st sess., H. Doc. 870, 1–7. See also Long, *Flood to Faucet*, 11–27, 59.

19. Daniel W. Mead, *Report on the Dam and Water Power Development at Austin, Texas*, 7–37; *Austin Statesman*, Apr. 23, 24, 25, 26, 1915; Jim Clayton interview with author, Austin, Texas, July 30, 1987. Massive rain and flooding in 1915 reached from mid-Oklahoma to the middle third of Texas. In Austin, rainfall exceeded 16.01 inches in the forty-eight hours from April 23 to early on April 25, 1915.

20. Colorado River Improvement Association, *Statement of Committee*, 1–2; U.S. Congress, House of Representatives, *Flood Control on the Colorado River, Texas*, 66th Cong., 1st sess., H. Doc. 304, 1–37; *United States Statutes at Large* (1916), 408–409; U.S. Department of Army, Chief of Engineers, *Preliminary Examination of Matagorda Bay, Texas*, 59th Cong., 1st sess., H. Doc. 154, 3–7.

21. U.S. Congress, House of Representatives, *Flood Control on the Colorado River*, 4.

22. Ibid., 12. Examples of the "most prominent" western projects were cited, and the Corps of Engineers recommended that a recent paper by Gen. H. M. Chittenden on "combination reservoirs" be consulted. See H. M. Chittenden, "Detention Reservoirs with Spillway Outlets As an Agency in Flood Control," *Proceedings of the American Society of Civil Engineers* (Sept., 1917): 1–23.

23. State of Texas, *Constitution*, 1876, art. 16, sec. 59.

24. *San Antonio Express*, Dec. 8, 1931; *Austin American*, Apr. 19, 1931.

25. U.S. Geological Survey, *The Floods in Central Texas in September, 1921*, by C. E. Ellsworth, Water Supply and Irrigation Paper no. 488, 1–14. On September 8–10, 1921, at least 224 lives were lost in south-central Texas.

26. Marc Reisner, *Cadillac Desert: The American West and Its Disappearing Water*, 115–50.

27. Ibid., 122–50; William G. Hoyt and Walter B. Langbein, *Floods*, 172–75, 183–330; "Memorandum in Re Muscle Shoals," Apr. 4, 1929, in Herbert C. Hoover Papers, Commerce Papers, box 199, Muscle Shoals, 1929, file; "Objections to the Norris Bill for Operation of Muscle Shoals," Nov. 24, 1930, in Hoover Papers, box 199, Muscle Shoals, 1929, file.

28. "Statement issued by Herbert Hoover, Secretary of Commerce and Major General Edgar Jadwin, Chief of Engineers, U.S.A., at Memphis, Tennessee, Apr. 30, 1927," in Hoover Papers, Commerce Papers, box 402, Mississippi Valley Flood Relief Work, Bible #725A; "Report to the President's Special Mississippi Flood Committee," May 2, 1927, ibid., Bible #730A; Hoyt and Langbein, *Floods*, 263–67, 370–71; Arthur E. Morgan, *Dams and Other Disasters: A Century of the Army Corps of Engineers in Civil Works*, 232–34.

29. "The Fight to Prevent Another Mississippi Flood," *The Literary Digest*, May 21, 1927, found in Hoover Papers, box 398, report file; "First Effects of the Mississippi Flood," *The Literary Digest*, June 18, 1927, ibid.; "Flood Demands Extra Session of Congress" (n.d.), in Hoover Papers, box 398, clippings file; Hoyt and Langbein, *Floods*, 173–76; Emmett Gloyna interview with author, Austin, Texas, Aug. 15, 1986; Pete Daniel, *Deep'n As It Come: The 1927 Mississippi River Flood*, 84–123. The most detailed account of the 1927 flood and reaction by Congress is "The Control of the Floods of the Mississippi River in the Alluvial Valley," by Mississippi representative W. H. Whittington, Mar. 5, 1928, *Congressional Record*, pp. 4246–54, found in Hoover Papers, box 403, Commerce Papers, legislative hearing file.

30. Herbert Hoover, "The National Policy of the Development of Water Resources," *Port and Terminal* 6 (Sept., 1926): 9.

31. Hoover to Coolidge, May 16, 1927, in Hoover Papers, Commerce Papers, box 407, relief work file; "Report to the President's Special Mississippi Flood Committee" (n.d.), ibid.; "Address of President Coolidge at the Seventh Annual Convention of the American National Red Cross," Oct. 3, 1927, ibid., box 408; Hoyt and Langbein, *Floods*, 370–71.

32. Richard Guy Wilson, "Machine-Age Iconography in the American West: The Design of Hoover Dam," *Pacific Historical Review* 54 (Nov., 1985): 468–93; Reisner, *Cadillac Desert*, 126–36; Richard Lowitt, *New Deal and the West*, 81–83; "Boulder Dam Legislation Urgent—Press Statement," Jan. 27,

1927, in Hoover Papers, box 81, Bible #70, Boulder Dam file; "Remarks of the President at Boulder Dam," Nov. 12, 1932, in Hoover Papers, box 105, Colorado-Hoover Dam file.

33. Forest McDonald, *Insull*, 221-29; Welly K. Hopkins interview, May 11, 1965, 18-20.

34. Hopkins interview, May 11, 1965; ibid., June 9, 1977.

35. John Williams, *The Story of the Lower Colorado River Authority*, 4; McDonald, *Insull*, 221-84.

36. Alvin J. Wirtz Papers, vertical file; Welly K. Hopkins interview by Joe B. Frantz, Nov. 14, 1968, 21-26; Hopkins interview, June 9, 1977.

37. Hopkins interview, Nov. 14, 1968.

38. Hopkins interview, May 11, 1965, and Nov. 14, 1968; *Austin American*, Apr. 18, 19, 1931. The Hamilton Dam was named for George W. Hamilton, a chief engineer with the Insull Company prior to its first withdrawal from the project in 1928. Hamilton became a consultant for both the TVA project and the Bonneville Power Administration in Washington. At the time a name was selected many wondered why it had not been named the Johnson Dam, after Civil War general and entrepreneur Adam Johnson. See also McDonald, *Insull*, 274-333.

39. James P. Buchanan Papers, vertical file; Wirtz Papers, vertical file; *Chicago Daily Tribune*, Sept. 25, 1934. See also G. C. Hill, "The History and Purpose of the Lower Colorado River Authority" (Ph.D. diss., University of Texas, Austin, 1935).

CHAPTER 2

1. Wirtz Biography, n.d., in Lyndon B. Johnson Library, Lower Colorado River Authority Papers, box PB 25, general information file; *Austin American*, Oct. 28, 29, 1951; *Dallas Morning News*, Jan. 3, 1940, May 24, 1941, Oct. 29, 1951; Welly K. Hopkins, interview, June 9, 1977.

2. P. W. Dent, "Texas Legislature in 1931," *The Reclamation Era* 23 (Feb., 1932): 24-26; Hopkins interview, June 9, 1977. See also Lyndon B. Johnson Papers, Selected Names, LBJA, boxes 36, 37, 38; Alvin J. Wirtz Papers, boxes 8, 9, 10, 11, 14; George Brown interview, July 11, 1977.

3. James P. Buchanan Papers, box 3J386, Appropriation Files and PWA Projects Record on Buchanan Dam.

4. Richard Lowitt, *New Deal and the West*, 218. See also Donald C. Swain, "The Bureau of Reclamation and the New Deal, 1933-1940," *Pacific Northwest Quarterly* 61 (July, 1970), 137-46; Walter Prescott Webb, *The Great Plains*, 348-66; Arthur Maass and Raymond L. Anderson, *. . . and the Desert Shall Rejoice: Conflict, Growth and Justice in the Arid Environments*.

5. J. Rupert Mason, "A New Deal for Reclamation," *The Reclamation Era* 25 (Feb., 1935): 25-26. As income from the sale of land diminished, the reclamation fund, after 1920, benefited from income generated from leasing oil rights in the public domain. However, the primary income produced after 1936 was revenue from sale of hydroelectric power.

6. Samuel P. Hays, *Conservation and the Gospel of Efficiency: The Progressive Conservation Movement, 1890-1920*, 100-101; Lowitt, *New Deal*

and the West, 81–90, 203–28; Everett H. Larson and David L. Goodman, "Reclamation Engineering," in *Dams and Control Works*, 1–11; Beverly B. Moeller, *Phil Swing and Boulder Dam*, 105–22. The original Reclamation Act of 1902 (32 Stat. 228) included only sixteen states. Texas was added in 1906, joining Arizona, California, Colorado, Oregon, South Dakota, Utah, Washington, Wyoming, Idaho, Kansas, New Mexico, North Dakota, Montana, Nebraska, Nevada, and Oklahoma.

    7. Donald J. Pisani, "Enterprise and Equity: A Critique of Western Water Law in the Nineteenth Century," *Western Historical Quarterly* 27 (Jan., 1987): 15–37; Norris Hundley, Jr., *Water and the West: The Colorado River Compact and the Politics of Water in the American West*, 1–73.

    8. Hays, *Conservation and the Gospel of Efficiency*, 101.

    9. Reisner, *Cadillac Desert*, 136. See also George O. Sanford, "Dams—High, Large, and Unusual, Part 1," *The Reclamation Era* 23 (Feb., 1932): 28; George O. Sanford, "Dams—High, Large, and Unusual, Part 2," *The Reclamation Era* 23 (Mar., 1932): 58–60. In 1932 there were only six major dams other than Hoover in the entire country that were completed or under construction: Wilson in Alabama, Bagnell in Missouri, Keokuk in Iowa, Conowingo in Maryland, Safe Harbor in Pennsylvania, and Rock Island in Washington. See John W. Haw and F. E. Schmitt, *Report on Federal Reclamation to the Secretary of the Interior*, 1–133.

    10. Albert U. Romasco, *The Poverty of Abundance*, 7; Leonard Tillotson, Gen. Manager Brazos River Conservation and Reclamation District to Hoover, Nov. 16, 1929.

    11. Herbert C. Hoover, *The Memoirs of Herbert Hoover: The Great Depression, 1929–1941*, 25–47.

    12. Romasco, *The Poverty of Abundance*, 125–234; Hoover, *Memoirs*, 46–156; Donald W. Whisenhunt, *The Depression in Texas: The Hoover Years*, 1–21, 134–49. New York was the first state to provide substantial relief funds to its citizens. In late 1931, at the request of Franklin D. Roosevelt that an emergency relief agency be established, the New York legislature unanimously approved $20 million for the Temporary Emergency Relief Administration (TERA). In early 1932 Harry L. Hopkins was named administrator and chair of TERA. See also Martin L. Fausold, *The Presidency of Herbert C. Hoover*, 63–166.

    13. Seth S. McKay and Odie B. Faulk, *Texas after Spindletop*, 132–37. As a political hot potato between the governor's office and the Texas legislature, the commission was alternately called the Texas Relief Commission, the Texas Rehabilitation and Relief Commission, the Texas Employment and Relief Commission, and the Texas Relief Commission Division; ultimately it became a part of the Division of Public Welfare.

    14. Lionel V. Patenaude, *Texans, Politics, and the New Deal*, 86–96.

    15. Westbrook to Hopkins, Dec. 21, 1933, in National Archives, Record Group 69, Records of the Work Projects Administration [hereafter NA, RG 69], box 44, Civil Works Administration [CWA] correspondence file.

    16. Davis to Hopkins, Dec. 16, 1933, in NA, RG 69, box 44, CWA Correspondence File, 1933; Westbrook to Hopkins, Jan. 11, 1933, ibid., box 45,

CWA Correspondence File; Westbrook to Hopkins, Nov. 29, 1933, ibid., box 44, CWA Correspondence File; Westbrook to Fellows, Jan. 11, 1934, ibid., box 45, CWA Correspondence File; Baker to Westbrook, Jan. 15, 1934, ibid.; Stone to Ferguson, Feb. 14, 1934, ibid.; Spillman to Ickes, Dec. 1, 1933, ibid., box 46, CWA Correspondence File.

17. Westbrook to Hopkins, Dec. 3, 1933, ibid., box 44, CWA Correspondence File.

18. Ibid. Westbrook's idea was carried out indirectly throughout the West. At most major reclamation projects entire communities were built at government expense to house and feed workers.

19. Westbrook to Williams, Dec. 28, 1933, ibid., box 282, Federal Emergency Relief Act [FERA] Correspondence File. Aubrey Williams had had the same firsthand experience as Westbrook in dealing with relief programs in Alabama prior to being brought to Washington by Harry Hopkins to head the rural relief program for the WPA. See George McJimsey, *Harry Hopkins: Ally of the Poor and Defender of Democracy,* 57-74, and James T. Patterson, *The New Deal and the States: Federalism in Transition,* 32-35.

20. Westbrook to Williams, Dec. 30, 1933, in NA, RG 69, box 282, FERA Correspondence File.

21. Patenaude, *Texans, Politics, and the New Deal.* 90-95.

22. Hopkins to ALL State Emergency Relief Administrations, Mar. 20, 1934, in Texas State Archives, Record Group 303, Records of State Purchasing and General Services Commission, Texas Relief Commission [hereafter TA, RG 303], Texas Relief Commission Correspondence.

23. FERA Manual Advance Bulletin Number 4 to ALL Emergency Relief Administrators and State Civil Works Administrations, March 27, ibid. See also Robert S. McElvaine, *The Great Depression in America, 1929-1941,* 264-70.

24. Davis to all County Administrators and Chairmen, Mar. 29, 1934, in TA, RG 303, Texas Relief Commission correspondence.

25. Patenaude, *Texans, Politics, and the New Deal,* 93-109.

26. Haw and Schmitt, *Report on Federal Reclamation,* 20-21; Hopkins to ALL State Emergency Relief Administrations, Mar. 20, 1934, in TA, RG 303, Texas Relief Commission Correspondence.

27. Haw and Schmitt, *Report on Federal Reclamation,* 20; *The Reclamation Era* 23 (Jan., 1932): 19; ibid., 24 (Apr., 1933): 41-42. For additional information see Kerwin J. Williams, *Grants-in-Aid under the Public Works Administration.* Williams concluded: "From the point of view of administration, the significant feature of the emergency public works program has been that the PWA relied primarily upon direct federal-local contracts, whereas previous federal grant agencies had confined themselves to dealing with the state government." See also Samuel I. Rosenman, ed., *The Public Papers and Addresses of Franklin D. Roosevelt,* vol. 4, 324.

28. Harold L. Ickes, "Our Right to Power," n.p.; Nov. 12, 1938, in Harold L. Ickes Papers [hereafter, Ickes Papers], box 105; H. Ickes, "In Defense of the New Deal Power Programs," n.p., Nov. 7, 1937, ibid.; H. Ickes, "Rivers of Strength in War and Peace," n.d., ibid., box 120. See also Graham White

and John Maze, *Harold Ickes of the New Deal: His Private Life and Public Career*, 142–96; *The Reclamation Era* 25 (1935): 209–10; Swain, "Bureau of Reclamation and the New Deal," *Pacific Northwest Quarterly* 61 (July, 1970): 137–45.

29. Swain, "Bureau of Reclamation and the New Deal," 138. See also William D. Reeves, "PWA and Competitive Administration in the New Deal," *Pacific Historical Review* 24 (Nov., 1965): 457; White and Maze, *Harold Ickes*, 114–20; Harry Hopkins and Harold Ickes to the President, "A National Work and Relief Program from July 1, 1935 to July 1, 1936," Oct. 6, 1934, in Ickes Papers, box 159, Federal Emergency Administration of Public Works File. During the first year of the New Deal Ickes had a fairly cooperative working relationship with Harry Hopkins. Yet, by late 1934 an intense rivalry developed between them over the use of funds and the selection of public works projects.

30. Ickes to Mead, Jan. 9, 1935, in National Archives, Bureau of Reclamation Records, Record Group 115 [hereafter, NA, RG 115], box 1096, File 301. Mead was not informed that Ickes was interested in doing a study and evaluation of the Colorado River. Mead, in early 1935, assumed that the "PWA project" in question concerned the Rio Grande.

31. Mead to Ickes, Jan. 12, 1935, ibid.; Ickes to Mead, Jan. 18, 1935; ibid.; Mitchell to Ickes, Feb. 12, 1935, ibid., box 1089, File 040.

32. P. W. Dent, "Texas Legislation in 1931," *The Reclamation Era* 23 (Feb., 1932): 24–26; Hopkins interview, June 9, 1977. Political interests were involved with the water rights, the partially completed Hamilton Dam was in receivership, the question of hydroelectric production and distribution was unresolved, and state administration of federal funding—already estimated to be in the millions of dollars—was being questioned.

33. State of Texas, *Senate Journal*, 43d Leg., 1st sess., Oct. 4, 1933, 171.

34. Ibid., 2d sess., Feb. 20, 1934, 174; Comer Clay, "The Lower Colorado River Authority: A Study in Politics and Public Administration" (Ph.D. diss., University of Texas, 1948), 89–90.

35. Federal Emergency Administration Release No. 804, 1934, in NA, RG 115, box 1094, Public Works Allotment File.

36. Ibid., *Austin American*, June 29, 30, 1934; *The Reclamation Era* 25 (Aug., 1935): 153; "Lower Colorado River Project, Texas," *The Reclamation Era* 25 (Sept., 1935): 187–90.

37. *Austin American*, July 18, 1934.

38. *Austin American*, Sept. 14, 1934; *Chicago Daily Tribune*, Sept. 25, 1934. An early case study of the private utilities companies' fear of federal or state inroads into their preserve is discussed in Moeller, *Phil Swing and Boulder Dam*.

39. Glavis to Administrator [Ickes], June 19, 1936, in Ickes Papers, box 149, Colorado River Authority File; Ickes to Glavis, July 16, 1936, ibid.; Hunt to Ickes, Sept. 18, 1936, ibid.; *Austin American*, Sept. 6, 1934; State of Texas, *Senate Journal*, 43d Leg., 3d sess., Sept. 17, 1934, 79.

40. *Dallas Morning News*, Sept. 23, 1934; *Austin American*, Sept. 23, 1934; State of Texas, *House Journal*, 43d Leg., 3d sess., Sept. 23, 1934, 331.

41. Hughes to Roosevelt, Oct. 3, 1934, Franklin D. Roosevelt Papers, New York Official File 114, Internal Waterways, box 1.

42. *Dallas Morning News*, Sept. 12, 1934; *Austin American*, Sept. 26, 1934; Clay, "Lower Colorado River Authority," 108–10; Ickes to Glavis, July 13, 1936, Ickes Papers, box 149, Colorado River Authority File.

43. Clay, "Lower Colorado River Authority," 105.

44. *San Angelo Evening Standard*, Oct. 8, 9, 10, 1934. Hays, in his *Conservation and the Gospel of Efficiency*, also points out the problem: "Plans to control rivers in the West had gone awry when *upstream* communities resisted water storage in their state for *downstream* use."

45. Clay, "Lower Colorado River Authority," 109; Thomas K. McCraw, *TVA and the Power Fight*, 1–25; Moeller, *Phil Swing and Boulder Dam*.

46. *San Angelo Standard Times*, Oct. 14, 15, 17, 1934; PWA Projects Record–Buchanan Dam Correspondence in Buchanan Papers, box 3J386 [hereafter, Buchanan Papers], 1934 File; Buchanan to Walter Woodward, Nov. 2, 1934, in Buchanan Papers; Woodward to Buchanan, Nov. 5, 1934, in Buchanan Papers. An Upper Colorado Authority was created by the Texas legislature in 1935. See also Kenneth B. Ragsdale, *The Year America Discovered Texas: Centennial '36*, 28–108.

47. *Austin Statesman*, Nov. 9, 1934; Buchanan to Woodward, Nov. 8, 1934, in Buchanan Papers.

48. Buchanan to Engelhard, Nov. 10, 1934, in Buchanan Papers.

49. Ickes to Hunt, Feb. 25, 1935, Ickes Papers, box 252, Jan–Feb 1935 File. See also Ickes to Hunt, Jan. 17, 1935, ibid.

50. Buchanan to Engelhard, Nov. 10, 1934, in Buchanan Papers, vertical file.

51. Mead to Hunt, Mar. 6, 1935, in NA, RG 115, box 1093.

52. Hunt to Buchanan, Dec. 3, 1934, in Buchanan Papers; Ickes to Buchanan, Dec. 1, 1934, ibid.; Buchanan to Ickes, Nov. 24, 1934, ibid.; Buchanan to Ickes, Dec. 4, 1934, ibid.; Smith to the Administrator [Ickes], Aug. 7, 1936, in Ickes Papers, box 149, Colorado River Authority, 1936, file. In early 1935, Governor Allred appointed Hughes as a judge of the Fourteenth District Court in Dallas. Nearly three decades after her fight to defeat the LCRA, Judge Sarah T. Hughes administered the oath of office to President Lyndon B. Johnson aboard Air Force One in Dallas on November 22, 1963.

53. Malott to Mitchell, Feb. 8, 1935, in Buchanan Papers; Mitchell to Ickes, Feb. 12, 1935, in NA, RG 115, box 1059, Texas Colorado River File 040; Mitchell to Mead, Mar. 4, 1935, ibid., box 1096, Engineering Reports File; Hunt to the Secretary [Ickes], Sept. 18, 1936, in Ickes Papers, box 149, Colorado River Authority.

54. Confidential: List No. 1004, Special Resolution Covering a Project of the Lower Colorado River Authority (Texas) for the Advisory Committee on Allotments, May 7, 1935, in NA, RG 115, box 1094, Public Works Allotments File 240.

55. Ibid.; Memorandum by H. T. Hunt Re: Lower Colorado River Authority (Texas), May 8, 1935, ibid.; Hunt to Ickes, May 6, 1935, ibid.; Roose-

velt to Ickes, May 28, 1935, ibid.; Roosevelt to Wallace, in Lyndon B. Johnson Library, July 2, 1935, House Papers, box 166, LCRA Buchanan Dam File. For a full discussion of the grants-in-aid program, revenue bonds, and funding of emergency aid by the PWA under the recovery act, see Williams, *Grants-in-Aid*, and Jack F. Isakoff, *The Public Works Administration*.

56. Press Release no. 3, Advisory Committee on Allotments, May 20, 1935, in NA, RG 115, box 1095, Public Works Allotments file; *Austin American*, May 17, 1935; "Lower Colorado River Project," *The Reclamation Era* 25 (Sept., 1935): 187–90; *Engineering News-Record* 114 (June 27, 1935): 928.

57. Harold L. Ickes, *Back to Work*, 226–28; idem, *The Autobiography of a Curmudgeon*, 293–99.

58. Ickes, *Back to Work*, 226–28.

CHAPTER 3

1. Lower Colorado River Authority, Minutes of the Board of Directors [hereafter, LCRA, Minutes], Feb. 19, 1935, 1–12.

2. *New York Times*, Dec. 31, 1933; LCRA, Minutes, Oct. 8, 1935. Initial allotments included $5 million for the Owyhee project; $2.7 million for the Deer Creek–Utah Lake project, $1.5 million for the Moon Lake Reservoir, $6 million for the All-American Canal, $4 million for the Verde River project, and $13 million for the Casper-Alcove project.

3. J. Rupert Mason, "A New Deal for Reclamation," *The Reclamation Era* 25 (Feb., 1935): 25–26; "One Hundred Million Dollars Set Aside for Reclamation," ibid. (Aug., 1935): 153; Roosevelt to Mr. Secretary [Ickes], July 2, 1935, in Lyndon B. Johnson Library, House Papers [hereafter, LBJ House Papers], box 166, LCRA Buchanan Dam File.

4. *New York Times*, Oct. 8, 1935; William G. Hoyt and Walter B. Langbein, *Floods*, 380–81; E. B. Debler and John Rider, *Marshall Ford Development Justification for High Dam*, 14, 18, in NA, RG 115, box 1098, Marshall Ford Dam and Reservoir, File 301.1

5. *Engineering News-Record* 115 (July 4, 1935): 29.

6. Ibid. See also Walter E. Long, *Flood to Faucet*, 93–95.

7. Ibid. Actual discharge at Austin was determined by the bureau to be 481,000 second-feet.

8. "Lower Colorado River Project, Texas," *The Reclamation Era* 25 (Sept., 1935): 190; Drew Pearson and Robert Allen, "The Washington Merry-Go-Round," *Washington Herald*, June 18, 1936; "Memorandum concerning purchase and properties from Colorado River Company and C. G. Malott and Settlement of Fegles Construction Company, Limited Content," May 11, 1936, in LBJ House Papers, box 166, Colorado River Authority File. The value as listed on the Lower Colorado River Authority balance sheet passed to Morrison, Malott, and Wirtz by the Central Texas Electric Company was $4,064,500. The court-appointed appraisers, led by John A. Norris, valued the property at $3,798,000, and the final sale price to the LCRA was $2,639,000. See also Glavis to Ickes, June 19, 1936, in Harold L. Ickes Papers, box 149, Colorado River Authority File; John E. Babcock interview, Nov. 22, 1983.

9. Hunt to the Secretary [Ickes], Sept. 18, 1936, in Ickes Papers, box

149, Colorado River Authority File; Glavis to Ickes, July 13, 1936, ibid.; Sim Gideon interview, Mar. 21, 1968. See also Malott to Mitchell, Feb. 8, 1935, in James P. Buchanan Papers, box 3J386, Buchanan Dam File.
   10. Schnupp to Walter, Aug. 30, 1935, in NA, RG 115, box 1089.
   11. *New York Times*, Oct. 8, 1935.
   12. "McDonough Gets CRA Project Job," n.d., in NA, RG 115, box 1088, Texas Colorado River File 023. See also LCRA, Board Minutes, 1935–36.
   13. Schnupp to Walter, Aug. 30, 1935, in NA, RG 115, box 1089.
   14. Elliott to Ickes, Oct. 12, 1935, ibid., box 1094, Texas Colorado River File 246; "Lower Colorado River Project, Texas," *The Reclamation Era* 25 (Sept., 1935): 187–90.
   15. Elliott to Ickes, Oct. 12, 1935, in NA, RG 115, box 1094, Texas Colorado River File 2460.
   16. "Era Allotments to Bureau Reduced by $20,000,000," *The Reclamation Era* 25 (Dec., 1935): 232–33; LCRA v. McCraw, 83 S.W. 2d 629 (1935).
   17. Elliott to Ickes, Oct. 12, 1935, in NA, RG 115, box 1094, Texas Colorado River File 246.
   18. Buchanan to Ickes, Oct. 14, 1935, ibid.
   19. Ibid.; McDonough to Ickes, Oct. 16, 1935, ibid.; Resolution of the Board of Directors of the Lower Colorado River Authority, undated, Doc. 16150, ibid., October 1935 file. All three of these documents use precisely the same language with regard to "specific purpose." Needless to say, Buchanan exercised a considerable influence over both the LCRA board and McDonough.
   20. Buchanan to Elwood Mead, Nov. 1, 1935, ibid., box 1096, Engineering Records File 301.
   21. Ibid.
   22. Ibid.
   23. Mead to Buchanan, Nov. 5, 1935, ibid.
   24. Bunger to Mead, Oct. 25, 1935, ibid.
   25. Mead to Buchanan, Nov. 5, 1935, ibid.
   26. Ibid.; McDonough to Ickes, Nov. 9, 1935, ibid.
   27. McDonough to Ickes, Nov. 9, 1935, ibid.; McDonough to Ickes, Nov. 12, 1935, ibid.; E. K. Burlew to Ickes, Nov. 12, 1935, ibid.; Ickes to Mead, Nov. 16, 1935, ibid.
   28. Mead to Ickes, Nov. 12, 1935, ibid. See also McDonough to Ickes, Nov. 13, 1935, ibid.
   29. Mead to Ickes, Nov. 19, 1935, ibid.
   30. Ickes to Mead, Nov. 25, 1935, ibid.
   31. Department of Interior, Press Release P.N. 110245, Dec. 10, 1935, ibid., box 1088, Texas Colorado River File 023.6.
   32. Mead to Ickes, Nov. 19, 1935, ibid., box 1096, Engineering Records File 301.
   33. Department of Interior, Press Release P.N. 111153, Jan. 3, 1936, ibid., box 1088, Texas Colorado River File 023.6; "Funds for Texas Project Restored," *The Reclamation Era* 26 (Feb., 1936): 42. Elwood Mead died in Jan., 1936, at the age of seventy-eight. He did not live to see the full realization of his efforts to protect and foster a broad multipurpose concept in the Bureau by the late

1930s. He did, however, have the future in mind when, in 1935, he brought John C. Page to Washington from the Hoover Dam project in order to groom him as his understudy and eventual successor. In Mead's honor Ickes named the vast body of water behind Hoover Dam, the largest manmade reservoir in the world, Lake Mead.

34. Roosevelt to Ickes, May 28, 1935, in NA, RG 115, box 1094, Public Works Allotments File. For further details see the following items: J. C. Page to Ickes, Mar. 19, 1936; Roosevelt to Ickes, June 24, 1935; R. N. Elliott to Ickes, Oct. 12, 1935; J. R. McCarl to Ickes, Nov. 29, 1935, ibid. J. R. McCarl, comptroller general of the United States quite possibly supplied the most crucial opinion when he stated in his Nov. 29, 1935, message to Ickes: "The allocation of $5,000,000 was legally obligated before the recession and, therefore, the $3,000,000 *would not*, as a result of the rescinding action become available for use for any other purpose, the proper adjustment at this time would appear to be restoration or reallocation of the $3,000,000 for use on this project." This conclusion supported Buchanan's claim that the funds to the LCRA were considered "special purpose" and thus untouchable except for the Colorado River project.

35. *Austin American*, Feb. 9, 1936; Department of Interior, Press Release, Feb. 28, 1936, in NA, RG 115, box 1088, Texas Colorado River File 023.6; "Colorado River Project, Texas," *The Reclamation Era* 26 (Mar., 1936): 71. The first major contract was for clearing fifteen thousand acres at the Buchanan Dam reservoir. Eleven companies bid, with the contract going to Brown and Root, Inc., of Austin, in the amount of $323,350.00. See also "Colorado River Project, Texas," *The Reclamation Era* 26 (Apr., 1936): 91.

36. *Austin American*, Feb. 20, Mar. 4, 1936; *Houston Post Sentinel*, Feb. 23, 1936.

37. "Colorado River Project, Texas," *The Reclamation Era* 26 (Apr., 1936): 84.

38. *Houston Post Sentinel*, Feb. 23, 1936; Kenneth E. Hendrickson, *The Waters of the Brazos: History of the Brazos River Authority, 1929–1979*, 21–49.

39. Samuel B. Hays, *Conservation and the Gospel of Efficiency: The Progressive Conservation Movement, 1890–1920*, 100–105, 114.

40. Ibid., 108–16; Ashwander et al. v. Tennessee Valley Authority et al., 297 U.S. 288 (1936).

41. Lilienthal to Roosevelt, Jan. 28, 1936, in Franklin D. Roosevelt Papers, box 3, TVA File: Attorney General to President, Feb. 17, 1936, ibid. David Lilienthal expressed grave concerns to Roosevelt that the court might rule against the TVA. In order to be prepared for a negative decision, Lilienthal had drafted a number of alternatives "left open to the government" in order to save the vast project.

42. *Austin American*, Mar. 6, 1936. For full details on suit see Reclamation Records, in NA, RG 115, box 1089, Community Public Service Company v. H. L. Ickes, Mar., 1936, Litigation File 070.

43. *Dallas Morning News*, Mar. 7, 1936.

44. *Austin American*, Mar. 6, 1936.

45. Ibid.

46. Ibid., Mar. 18, 1936. The Brazos project, although appropriated $30 million, had only received about $1.5 million to begin work on a project at Possum Kingdom in Young and Palo Pinto counties. In the case filed in the District of Columbia, the private utilities described the flood-control and irrigation classifications as "shams."

47. Ibid., Mar. 7, 1936; Malott to Buchanan, Apr. 17, 1936, in Buchanan Papers, vertical file.

48. *Austin American*, Mar. 6, 1936; *Houston Chronicle*, Mar. 7, 1936; *Dallas Morning News*, Mar. 7, 1936; *San Antonio Express*, Mar. 7, 1936. See also Babcock interview.

49. *Austin American*, Mar. 6, 1936.

50. Ibid., Mar. 7, 1936.

51. Ibid., Mar. 9, 1936. Fielding Hammond of Burnet commented, "If we'd go back to the burning of oil lamps as in olden days it would go far in teaching the utilities that the US Government should have the right to assist the common view of people." And this statement came from H. T. Harrison, a Llano automobile dealer: "My opinion is that the utilities companies are not justified in taking the stand against the bureau of reclamation and are attempting to hinder progress of developing the natural resources and depriving the people especially in the smaller towns and rural communities of modern household conveniences."

52. Ibid., Mar. 18, 1936.

53. Ibid., Mar. 11, 12, 1936; *San Antonio Express*, Mar. 15, 1936.

54. *Austin American*, Mar. 15, 1936. The federal court released $175,000 for supplies in order to prevent delays at the Buchanan site.

55. Ibid., Mar. 13, 1936.

56. Ibid., Mar. 14, 18, 1936.

57. Kubach to Page, Mar. 30, 1936, in NA, RG 115, box 1094, Public Works Allotments File.

58. Ibid.

59. Ibid. See also Burlew to Page, Mar. 25, 1936, ibid.

60. *Austin American*, Apr. 4, 9, 1936.

61. Department of Interior, Press Release, Apr. 9, 1936, in NA, RG 115, box 1088, Texas Colorado River File 023.6; "Colorado River Project, Texas," *The Reclamation Era* 26 (May, 1936): 117; Bobby Bostic interview with author, Buchanan Dam, Texas, Apr. 20, 1987. In another matter, Brown and Root, Inc., was awarded a contract on April 13, 1936, in the amount of $38,507 for construction of the north dike at the Buchanan reservoir.

62. Ickes to Page, Apr. 18, 1936, in NA, RG 115, box 1096, Engineering Reports File; Page to Bunger, Apr. 20, 1936, ibid.

63. Bunger to Page, June 3, 1936, ibid., Texas Colorado River File 301.

64. Ibid.

65. Page to Harper, June 6, 1936, ibid., Engineering Reports File.

66. Ibid. See also Department of Interior, Press Release, June 22, 1936, ibid.

67. Page to Harper, June 6, 1936, ibid.

68. Ibid.

69. Ibid. Grand Coulee Dam would be funded separately in the Interior Bill for $20 million, and the key dam in the Central Valley project, Friant Dam, would receive an additional appropriation of $6.9 million to augment an allocation of $8.1 million for canals and superstructures.

70. Ibid.

71. Ibid. In the weeks following the agreement between the LCRA and the bureau, Buchanan boasted in the company of the bureau representative that "the Bureau was so 'damn' slow on the Buchanan Dam that he took the work away from them and placed it so that it could be speeded up." See Report on Meetings with the Representatives of the Lower Colorado Flood Control Association, ibid., box 1095, Engineering Reports File.

72. *Austin American*, June 18, 1936; *San Antonio Express*, June 21, 1936; Department of Interior, Press Release, June 22, 1936, in NA, RG 115, box 1096, Engineering Reports File; Bunger to Commissioner of Reclamation, July 2, 1936, ibid., box 1095, File 301. Alvin Wirtz remained in Washington; McDonough and Fry returned to Texas to supervise the heavy construction. Wirtz remained primarily to ensure the speedy and proper transfer of funds back to the LCRA. While in Washington, he made numerous acquaintances that would be of future importance. See also "Colorado River Project, Texas," *The Reclamation Era* 26 (Aug., 1936): 186.

73. Walter to Page, June 23, 1936, in NA, RG 115, box 1096, File 301. Bunger knew as early as June 13 that the bureau would relinquish its rights to the LCRA. However, on this date the final contracts had not been signed. See also Bunger to Walter, June 13, 1936, ibid., Engineering Reports File; *Austin American*, June 24, 1936; *San Antonio Express*, June 28, 1936.

74. Walter to Page, June 23, 1936, in NA, RG 115, box 1095, Engineering Reports—Confidential Files, File 301.

75. Bunger to Walter, June 15, 1936, ibid. The Colorado River watershed again flooded in mid-September, 1936, even though the precipitation was concentrated in the upper river above San Angelo. The 26,350 square miles of watershed above Austin had an average rainfall of 7.5 inches between September 14 and 17. The discharge due to the large area covered had an exceptionally low peak rate of discharge of 150,000 second-feet. The floods of 1935 and 1936 provided more data on flooding characteristics along the Colorado than all the previous information combined.

76. Bunger to Walter, June 17, 1936, ibid.

77. Walter to Page, June 23, 1936, ibid.

78. Ibid.

79. Page to Walter, June 25, 1936, ibid.

80. Ibid.

81. The last report of cracks was made by Bunger to the Denver office on July 21, 1936, one month after the LCRA took over the Buchanan site. It is assumed that a detailed inspection report was made available to the authority. See Bunger to Chief Engineer, Denver, and Report on Cracks, July 21, 1936, ibid., Engineering Reports File; Walter to Page July 31, 1936, ibid.

82. *San Antonio Express*, July 6, 1936. This was the second major flood below Austin in the first half of 1936. Bunger, who was still being hounded

by McDonough and the Lower Colorado Flood Control Association to conduct a survey south of Austin (which he refused), reported that flooding at Matagorda was five inches short of being as high as the major 1935 flood. Bunger also made an important observation: "These floods [1936] have proven that the upper dam alone will not give any degree of protection to the vast territory below Columbus, and, since the loss of life and property damage for the Colorado River floods is mainly in the lower end, no flood program would be complete unless protection is given to this territory." See NA, RG 115, box 1095, Engineering Reports File.

83. *Austin American*, July 7, 1936; "Colorado River Project, Texas," *The Reclamation Era* 26 (June, 1936): 145. In order to provide supplemented power to the Buchanan site, workers began in July to repair the flood-damaged Marble Falls power plant. The action was intended to reduce dependence upon Texas Power and Light, from which the LCRA was buying about $800 worth of power per month.

84. *Austin American*, July 15, 16, 19, 24, Nov. 22, Dec. 18, 1936; Hoyt and Langbein, *Floods*, 383.

85. *Austin American*, Dec. 2, 18, 1936; Federal Emergency Administration of Public Works, Press Release no. 3271, n.d., in Lyndon B. Johnson, House Papers, box 166, Colorado River Authority File.

86. *Austin American*, Jan. 8. 1937; "General Information Concerning the Colorado River Project, Texas," July 1, 1938, in NA, RG 115, box 1102, Correspondence re Power Transmission Lines File; "Resources of Colorado River Dams," n.d., in Lyndon B. Johnson House Papers, box 166, Colorado River Authority File. By Christmas, 1936, over twenty-five hundred workers were employed at the Buchanan and Inks sites.

87. *Austin American*, Dec. 2, 18, 1936; Thomas K. McCraw, *TVA and the Power Fight*, 120; Duke Power et al. v. Greenwood County et al., 302, U.S. 485 (1938); Babcock interview, Nov. 22, 1983. See also Joseph C. Swidler and Robert H. Marquis, "TVA in Court: A Study of TVA's Constitutional Litigation," *Iowa Law Review* 32 (Jan., 1947): 296–326; George D. Haimbaugh, Jr., "The TVA Cases: A Quarter Century Later," *Indiana Law Journal* 41 (Winter, 1966): 197–227.

88. *Austin American*, Dec. 19, 20, 1936; *Washington Star*, Dec. 20, 1936; Federal Emergency Relief Act (FERA), Press Release no. 3271, n.d., in Lyndon B. Johnson House Papers, box 166, Colorado River Authority File. The suit by Duke was to prevent the advancement of $2.5 million to Greenwood County, and was first filed in federal court on Nov. 7, 1934. For a good review of the legal fight between the private sector utilities and the public agencies funded by the PWA, see McCraw, *TVA and the Power Fight*, 108–21.

89. "Invitation to Bidders on Generators and Turbines," n.d., in NA, RG 115, box 1088, Press Clippings File 023. The initial contract for generators at the Buchanan powerhouse totaled over $700,000.

90. FERA, Press Release no. 3271, n.d., in Lyndon B. Johnson House Papers, box 166, Colorado River Authority File. See also "Our Right to Power," Nov. 12, 1936, in Ickes Papers, box 105, 1938.

91. Duke Power Co. et al. v. Greenwood County et al., 302 U.S. 485

(1938); Alabama Power Company v. Ickes et al. 302 U.S. 464 (1938). Both of these cases were decided on Jan. 3, 1938.

92. Ickes to Page, Feb. 27, 1937, in NA, RG 115, box 1098, Marshall Ford Dam and Reservoir File.

CHAPTER 4

1. Merle Miller, *Lyndon: An Oral Biography*, 57–61; Robert A. Caro, *The Years of Lyndon Johnson: The Path to Power*, 380–441; *Congressional Record*, Appendix, 75th Cong., 1st sess., May 26, 1937, 1289–90; ibid., Mar. 3, 1937, 801–804. In addition to his substantially aiding the LCRA project, Buchanan was instrumental in securing funding for both the Central Valley water project in California and the Chickamauga Dam on the Tennessee River near Chattanooga.

2. *Austin American*, Feb. 28, 1937; "Four Large Dams to Harness Power," *San Antonio Express*, Feb. 17, 1937, found in NA, RG 115, box 1088, File 023; G. W. Morrison, "The Buchanan Dam," *Compressed Air Magazine* 41 (Nov., 1936): 5157–63; Walter K. M. Slavick, "Monuments to the Living," *The Reclamation Era* 30 (Feb., 1940): 42–45. In order of their size, the five largest dams in the world in 1940 were Grand Coulee, Shasta, Hoover, Friant, and Marshall Ford. Hoover is the highest at 726 feet, yet Grand Coulee contains three times as many cubic yards of construction material as Hoover and twice as many as Shasta.

3. The Marshall Ford Dam is named for a rancher, Ed Marshall, who leased the property near Horseshoe Bend, where the dam was to be built. The location had been referred to as the Hughes site or the Maxwell site, names of the property owners. Once the final location was chosen, it was the intention of the bureau and the LCRA to refer to it as the Marshall Ford Dam or simply Marshall Ford. See *San Antonio Express*, Feb. 17, 1937.

4. Buchanan to Burlew, Aug. 14, 1936, in NA, RG 115, box 1095, Engineering Reports File 301.

5. Ibid.; *The Austin Review*, July 23, 1936; Construction Engineer to Chief Engineer, Aug. 1, 1936, in NA, RG 115, box 1089.

6. Press Release, Dec. 4, 1936, in NA, RG 115, box 1088, Texas Colorado River File 023.6; *San Antonio Express*, Dec. 5, 1936; *Austin American Statesman*, Nov. 22, 1936; *The Reclamation Era* 27 (Jan., 1937): 18, and 26 (Nov., 1936): 257; *Engineering News-Record* 117 (Oct. 29, 1936): 626, and 117 (Dec. 10, 1936): 844.

7. Caro, *Years of Lyndon Johnson*, 269–80; John E. Babcock interview, Nov. 22, 1983; George Brown interview, Apr. 6, 1968, Aug. 9, 1969. For a full discussion of the activities of George and Herman Brown see Caro's chapter "The Dam." Wirtz most notably helped avert a delay in funding and construction by finding a solution to a deed dispute at the Marshall Ford site. The location was supposed to be owned by the federal government prior to construction, but due to an oversight it was discovered in late 1936 that Texas, when it joined the Union, had full ownership of all its public lands, including riverbeds. Wirtz and Buchanan resolved the problem. For partial information on the compensation Wirtz received, see Lower Colorado River Authority

Papers, PB 27, General Information File and Wirtz Compensation, 1935–1939.
    8. *Austin American*, Jan. 24, Feb. 3, 1937; *The Reclamation Era* 27 (Feb., 1937): 44, and 27 (Aug. 1937): 195. In addition to Whipple, Ceylon P. Humphreys and Arthur L. Gray, inspectors on the Ogden River project, and Homer H. Mills, inspector on the Boulder Canyon project, were transferred to the Texas project.
    9. Press Release, P.N. 130253, Feb. 2, 1937, in NA, RG 115, box 1088, File 023.6.
    10. *Austin American*, Jan. 24, Feb. 7, 1937; *Austin Statesman*, Feb. 16, 18, 1937.
    11. *San Antonio Express*, Feb. 17, 1937; *Houston Chronicle*, Feb. 17, 1937; "Ickes Defends PWA Program before Texas," Feb. 18, 1937, in NA, RG 115, box 1088, Texas Colorado River File 023; "3 Years of PWA," *The Reclamation Era* 27 (Mar., 1937): i. See also George McJimsey, *Harry Hopkins: Ally of the Poor and Defender of Democracy*.
    12. *San Antonio Express*, Feb. 17, 1937; *Austin American*, Feb. 14, 1937; *Dallas Morning News*, Feb. 20, 1937.
    13. *Austin Statesman*, Feb. 18, 19, 1937. Prior to the dedication of the Marshall Ford, Ickes, accompanied by Gov. James V. Allred, Lt. Gov. Walter Woodul, A. J. Wirtz, and LCRA officials, inspected and walked the entire length along the top of the 10,500-foot Buchanan Dam. See also "Dedication of Marshall Ford Dam, Colorado River Project, Texas," *The Reclamation Era* 27 (Apr., 1937): 69–71.
    14. "Dedication of Marshall Ford Dam, Hon. Harold L. Ickes," Feb. 19, 1937, in NA, RG 115, box 1080, File 090.20.
    15. Ibid.; *El Paso Times*, Feb. 20, 1937.
    16. *Dallas Morning News*, Feb. 20, 1937; *Austin Statesman*, Feb. 23, 1937; *Washington Post*, Feb. 23, 1937; *New York Times*, Feb. 23, 1937. The *Washington Post* praised Buchanan, noting "he piloted through the chamber the $4,880,000,000 work relief appropriation, the largest single money measure in the country's history." However, Roosevelt's court-packing scheme was in full debate, overshadowing the congressman's death.
    17. *Austin American*, Feb. 28, 1937; Caro, *Years of Lyndon Johnson*, 389–401; Anthony M. Orum, "Enter Lyndon Johnson," *The Texas Observer* 77 (Jan. 11, 1985): 13–15; *Austin Statesman*, Mar. 1, 1937.
    18. "FDR Court Plan to Get First Test: Lyndon Johnson Submits Issue to People," *Austin Statesman*, Mar. 6, 1937. Johnson's reference to the federal court in Washington, D.C., refers to the numerous suits bought by power companies against the PWA and secretary Ickes. Wirtz throughout this period was active in representing the LCRA's interest as well as other defendants in these suits. See also Babcock interview, Nov. 22, 1983.
    19. Caro, *Years of Lyndon Johnson*, 394–447; *Austin Statesman*, Mar. 1, 2, 1937; Sim Gideon interview, Mar. 21, 1968, 12–13. Wirtz would be a lifelong advisor to Johnson. He chaired the National Youth Administration Advisory Board in late 1936 in Texas and was Johnson's legal counsel and key political strategist. See also Llewellyn B. Griffith, Sr. interview, Aug. 15, 1978; Welly K. Hopkins interview, June 9, 1977; George Brown interview, Apr. 6, 1968.

20. Caro, *Years of Lyndon Johnson*, 448–49; Roosevelt's speech, May 11, 1937, College Station, Texas, in Franklin D. Roosevelt Papers, President's Personal File 1053; D. B. Hardeman and Donald C. Bacon, *Rayburn: A Biography*, 230–42; Oscar Chapman interview, Feb. 5, 1981; Gordon Fulcher interview, Jan. 13, 1969.
21. *Austin American*, Feb. 4, 1937.
22. Ibid., Apr. 30, 1937.
23. Press Release, no. 3139, May 3, 1937, in NA, RG 115, box 1088, File 023.6; *San Antonio News*, May 4, 1937; *Austin American*, May 10, 1937.
24. Page to Mansfield, May 5, 1937, in NA, RG 115, box 1098, Marshall Ford Dam and Reservoir File 301.1. See also "Towers Used on Topographic Survey of the Marshall Ford Reservoir Site, Colorado River Project, Texas," *The Reclamation Era*, 27 (May, 1937): 108–109, 118.
25. Ibid. AF stands for acre-feet of storage.
26. Marshall Ford Dam, updated press release by Congressman Johnson, May 20, 1937, in NA, RG 115, box 1098, File 301.1.
27. *Austin American*, Apr. 8, 1937.
28. Ibid., Apr. 15, 1937.
29. Ibid., Apr. 13, 1937. Taylor was a member of the Citizen's Advisory Committee in 1937 that advised the Austin City Council to rebuild the dam.
30. Ibid., Apr. 30, 1937.
31. "Council Studies Dam Completion," n.d., in NA, RG 115, Press Clippings File 023; Apr. 30, 1937.
32. *Austin American*, Apr. 30, 1937. The Bureau of Reclamation did not act on the proposal. There was already a dam in Arizona named Roosevelt, after Theodore Roosevelt.
33. *Austin American*, May 15, 22, 1937.
34. Ibid.
35. Ibid., May 15, 18, 20, 22, 1937.
36. Ibid., May 27, 1937.
37. *Austin American*, May 27, 28, 29, June 3, 27, 29, 1937. Johnson later noted that Ickes expressed deep concern over the controversy surrounding the Austin Dam. Ickes felt that negotiations should have been conducted in Austin and not in Washington. See Johnson to Wirtz, July 12, 1937, in Alvin J. Wirtz Papers, box 12. Also, the LCRA agreed to designate $200,000 for immediate site work and even made plans to move idle equipment from the Buchanan Dam to the Austin site.
38. *Austin American*, June 3, 4, 5, 30, 1937; "A New Dam Goes into Service," *Engineering News-Record* 119 (July, 1937): 39; *The Reclamation Era*, 27 (Aug., 1937): 183; "Marshall Ford Dam," *Construction Methods and Equipment* 19 (June, 1937): 64–65.
39. R. L. Lowry, "Flood Control by Marshall Ford Reservoir," May 1937, in NA, RG 115, box 110, File 301.4; E. B. Debler and John R. Rider, "Marshall Ford Development: Justification for High Dam," Aug., 1937, ibid., box 1098, File 301.1.
40. *Austin American*, June 24, July 2, 1937.

41. Johnson to Wirtz, July 12, 1937, in Wirtz Papers, box 12, Wirtz Correspondence File.

42. William Leuchtenburg, "Franklin D. Roosevelt's Supreme Court Packing Plan," in Wilmon H. Droze, George Wolfskill, and William E. Leuchtenburg, *Essays on the New Deal*, 88–89; James T. Patterson, *Congressional Conservatism and the New Deal: The Growth of the Conservative Coalition in Congress, 1933–1939*, 77–125.

43. William E. Leuchtenburg, *Franklin D. Roosevelt and the New Deal, 1932–1940*, 242–67; Kenneth D. Roose, "The Recession of 1937–38," *The Journal of Political Economy* 56 (June, 1948): 239–48; John Morton Blum, *From the Morgenthau Diaries*, vol. 1, 393.

44. Leuchtenburg, *Roosevelt and the New Deal*, 259–61, 263; John Major, *The New Deal*, 139–57, 244–46. The "Roosevelt Recession" wiped out over two years of economic progress. Numbers on unemployment rolls shot from 7.7 million in early 1937 to over 10.4 million by late 1938. All major economic indicators dropped; the primary ones were the index of manufactured products down 23 percent, farm price index down 20.5 percent, wholesale price index down 9 percent and the gross national product (GNP) down .05 percent. Steel production dropped to 19.2 percent of capacity in December, 1937.

45. Patterson, *Congressional Conservatism and the New Deal*, vii–ix, 74–287; Blum, *Morgenthau Diaries*, vol. 1, 393; Major, *New Deal*, 139–49; "The New Deal in Review, 1936–1940," *The New Republic* 47 (May, 1940): 705–708; Leuchtenburg, *Roosevelt and the New Deal*, 346–48; Samuel L. Rosenman, ed. *The Public Papers and Addresses of Franklin D. Roosevelt*, vol. 7, 33–34, 221–22, 225–26.

46. Johnson to Wirtz, July 12, 1937, in Wirtz Papers, box 12, Wirtz Correspondence File.

47. Press Release, July 30, 1937, in NA, RG 115, box 1088, File 023.6; Burlew to Page, July 31, 1937, ibid., box 1094, Public Works Allotment File. The president approved the $5 million to come from the Emergency Relief Appropriations Act of 1937. See also *Calexico Chronicle*, Aug. 2, 1937, in NA, RG 115, box 1088, File 023, and "News of the Week," *Engineering News-Record* 119 (Aug., 1937): 7.

48. Ickes to Page, Nov. 20, 1937, in NA, RG 115, box 1095, File 301.1. Debate on the status of Marshall Ford was handled at the highest level in Congress and the White House. Cost estimates from the LCRA and the bureau were constantly "in conflict with each other and with themselves." See also Marshall Ford Dam Revised Contract Estimate, Dec. 20, 1937, ibid., box 1098; Mansfield to Page, Dec. 16, 1937, ibid., box 1098, File 301.1; Barber to Mansfield, Dec. 6, 1937, ibid.; LeTulle to Mansfield, Dec. 16, 1937, ibid.

49. "Address by Secretary of the Interior and Federal Administrator of Public Works Hon. Harold L. Ickes at the Dedication of the Buchanan and Inks Dam, Texas, Saturday, October 16," Oct. 16, 1937, in Lower Colorado River Authority Papers [hereafter, LCRA Papers], box 28; Caro, *Years of Lyndon Johnson*, 459–62. See also Orum, "Enter Lyndon Johnson," *The Texas Observer* (Jan. 11, 1985): 13–16. Both sources provide a full discussion of the relation-

ship between George Brown of the Brown and Root Company, Tom Miller, and Alvin Wirtz with Johnson. Orum concludes that Johnson was under tremendous pressure to secure federal funds for the dams.

50. Bunger to Walter, Nov. 5, 1937, in NA, RG 115, box 1095, File 301; In order to make maximum use of the engineering talent available, the Interior Department transferred Ernest A. Moritz, a specialist in concrete dams, from Parker Dam in California to Marshall Ford. Howard P. Bunger, who had cooperated with the Brown and Root company at Marshall Ford during the initial phases of construction, was sent to Parker Dam to install the floodgates and bring the dam to completion. Parker Dam is a concrete dam which is very similar to, yet smaller than, Buchanan. See Press Release P.N. 12168, Dec. 29, 1937, ibid., box 1088, Texas Colorado River Authority File 023.6.

51. Press Release P.N. 14632, Feb. 2, 1938, in NA, RG 115, box 1088, File 023.6; Press Release P.N. 17592, Mar. 7, 1938, ibid., box 1088, Correspondence Re Purchase of Cement and Concrete File; Page to Johnson, Jan. 10, 1938, ibid., box 1098, File 301.1.

52. Ibid.

53. Page to Johnson, Feb. 26, 1938, ibid., box 1098, File 301.1. See also Caro, *Years of Lyndon Johnson*, 466.

54. Page to Johnson, Feb. 26, 1938, in NA, RG 115, box 1098, File 301.1.

55. Ickes to Johnson, Mar. 1, 1938, ibid.

56. Ibid., Page to Johnson, Mar. 2, 1938, ibid.

57. Moritz to Page, May 21, 1938, ibid., box 1088, File 23.6; Press Release, May 31, 1938, ibid., box 1098, File 301.1.

58. Moritz to Chief Engineer, Apr. 29, 1938, ibid., box 1088, File 023; *Austin Statesman*, Apr. 28, 1938; Wirtz to Johnson, in Lyndon B. Johnson, House of Representatives Papers, box 176, Lower Colorado River Authority File.

59. *Austin Statesman*, June 6, 1938, found in NA, RG 115, box 1088, File 023.

60. Wirtz to Moritz, June 16, 1938, ibid., box 1098, File 301.1; Moritz to Wirtz, June 17, 1938, ibid.; Wirtz to Burlew, June 24, 1938, ibid.; Burlew to Wirtz, July 1, 1938, ibid. During the negotiation to gain more funding Wirtz used Johnson's name and congressional office, even though not a paid employee of the Johnson staff, in order to persuade both the Interior Department and the Reclamation Bureau.

61. Little did Wirtz and Johnson know at the time of the July–August, 1938, Colorado River flood that the president, under the provisions of the Public Works Administration Appropriations Act of 1938, did allot $1.25 million for "widening the foundation at Marshall Ford." Their three-part strategy had paid off. See Roosevelt to the Administrator [Ickes], n.d., ibid., box 1094, File 246; Ickes to Secretary of the Treasury, July 30, 1938, ibid.; Burlew to Page, Aug. 1, 1983, ibid., Press Release no. 3360, n.d., ibid., box 1088, File 023.6.

62. En route by train to the West Coast, Roosevelt stopped in Amarillo, Texas, on July 11, 1938, to make a speech—in a heavy rainstorm—in which he made a polite reference to Lyndon Johnson "down in Austin . . . keeping his land from washing away." The original stenographic transcript of his address is water-stained from the storm. The president signed the last page of

the address: "Franklin D. Roosevelt (Original—delivered in a downpour)." See Franklin D. Roosevelt Speech, July 11, 1938, Amarillo, Texas, in Roosevelt Papers, President's Personal File 1156.
  63. *Wall Street Journal*, July 26, 1938; *Austin Statesman*, June 6, July 27, 1938.
  64. McKenzie to Page, July 27, 1938, in NA, RG 115, box 1100, Texas Colorado River Flood Control File.
  65. McKenzie to Page, July 28, 1938, ibid. McKenzie and the Brown and Root Company had pressured both Johnson and the Interior Department to extend their contracts. See Caro, *Years of Lyndon Johnson*, 459–68.
  66. *San Antonio Express*, July 28, 1938.
  67. Ibid. See also telegram from E. O. Taulee, President—Lower Colorado River Flood Control Association, to Page, Aug. 3, 1938, in NA, RG 115, box 1100, Texas Colorado River Flood Control File.
  68. *San Antonio Express*, July 28, 1938.
  69. McKenzie to Page, July 29, 1938, in NA, RG 115, box 1100, Texas Colorado Flood Control File.
  70. "Blunders at Buchanan Dam Blamed for Flood Disaster," *Arizona Republic*, Aug. 10, 1938; "PWA Denies Dam Caused Flood Damage," ibid., Aug. 12, 1938; "PWA Declares PWA Dam Not Guilty in Texas Flood," *Washington Post*, Aug. 11, 1938; "Hydromania in Texas," *Los Angeles Times*, Aug. 13, 1938; "Full Bucket," *Time* 37 (Aug. 8, 1938): 11–12, found in Reclamation Records, in NA, RG 115, box 1088, Clippings File, and box 1100, Texas Colorado River Flood Control File.
  71. "Full Bucket," *Time*, 37 (Aug. 8, 1938): 11–12, in NA, RG 115, box 1088, Clippings File. See also "Flood Making (Control) Dam in Texas," *Time*, n.d., ibid., box 1100, Texas Colorado River Flood Control File.
  72. *Washington Post*, Aug. 11, 1938.
  73. Ibid.
  74. Hampton to Ickes, Aug. 12, 1938, in NA, RG 115, box 1100, Texas Colorado River Flood Control File; Hawes to Page, Aug. 13, 1938, ibid., box 1098, File 301.1.
  75. Ibid. For a conflicting view opposed to the LCRA power production see McKinney to Roosevelt, Aug. 4, 1938, and attached article "WTU [West Texas Utilities] Company Says Dam Cannot Serve 2 Purposes," ibid., box 1100, Texas Colorado River Flood Control File. These sentiments attacked Roosevelt.
  76. "Statement of Lyndon B. Johnson to LCRA Board," Apr. 12, 1938, in LCRA Papers, PB 26, General Information, U.S. Government, Lyndon B. Johnson 1937–1938 File.
  77. Press Release no. 3425, n.d., in NA, RG 115, box 1088, File 023.6. Bashore had been with the bureau since 1906 and had taken a temporary leave from his position as chief construction engineer at the Kendrick project, Wyoming (Seminoe and Alcora dams) to conduct the investigation of the Texas flood. See also *Engineering News-Record* 121 (Aug. 18, 1938): 187.
  78. "Colorado River Flood, July–August 1938: Report of the State Board of Water Engineers," Sept. 19, 1938, in NA, RG 115, box 1100, Colorado River

Flood, July–August 1938 File; Abraham Streiff, *Technical Memorandum on the Colorado River Flood of July, 1938*, Aug. 23, 1938, in LCRA Papers, PB 71, General Information, Flood 1938 File.

79. Excerpts from the Testimony of C. W. Hackett, n.d., in LCRA papers, PB 72, C. W. Hackett Testimony File.

80. Summary of Complaints, n.d., ibid., PB 71, General Information, Flood 1938 File.

81. Ibid. See also LCRA v. Gulf Coast Water Co., 107 S. W. 2d 1101 (1937).

82. Report of Senate Investigating Committee, *Concerning Colorado River Authority*, n.d., in LCRA Papers, PB 71, General Information, Flood 1938 File; *Engineering News-Record* 121 (Sept. 29, 1938): 388; Kendrick A. Clements, "Committee Disapproves Reservoir Level Law," *Engineering News-Record* 122 (Mar. 2, 1939): 257.

83. Testimony, n.d., in LCRA Papers, PB 72, Testimony File.

CHAPTER 5

1. Statement of Lyndon B. Johnson, Apr. 12, 1938, in Lower Colorado River Authority Papers [hereafter, LCRA Papers], box PB 26, General Information, U.S. Government, Lyndon B. Johnson, 1937–1938 File; *New York Times*, July 31, 1938; *Washington Post*, July 10, 1938; *Wall Street Journal*, July 27, 1938.

2. Kenneth B. Keener, "The Low Dam at Marshall Ford," *Engineering News-Record* 121 (Dec. 1, 1938): 697–99; Draft of Material for Press Release, Doc. 85138, Dec. 3, 1938, in NA, RG 115, box 1088, Press Release File; Moritz to Chief Engineer Walter, Dec. 5, 1938, in NA, RG 115, box 1098, File 301.1.

3. Johnson to Ickes, Feb. 17, 1939, in NA, RG 115, box 1098, File 301.1.

4. "Multi-Purpose Dams Opposed in Texas," *Engineering News-Record* 122 (Feb. 9, 1936): 6; *Austin American*, Aug. 8, 1938; *Washington Post*, Aug. 12, 1938.

5. Kendrick A. Clements, "Committee Disapproves Reservoir Level Law," *Engineering News-Record* 122 (Mar. 2, 1939): 257; "LCRA Announces Flood Policy," *Engineering News-Record* 122 (Mar. 30, 1939) found in NA, RG 115, box 1098, File 301.1; Page to Johnson, Mar. 22, 1939, in NA, RG 115, box 1098, File 301.1; John C. Page, "The Multiple-Purpose Project," *The Reclamation Era* 29 (May, 1939): 93–95.

6. Page to Johnson, Mar. 27, 1939, in NA, RG 115, box 1098, File 301.1; Burlew, First Assistant Secretary and Budget Officer [Interior], to the Acting Director, Bureau of the Budget, Apr. 3, 1939, ibid., box 1099, Flood, Levees, Dikes File 301.4; *The Reclamation Era* 29 (Apr. 1939): cover; Harry Provence, *Lyndon B. Johnson*, 35–51. See also Caro, *Years of Lyndon Johnson*, 469–605. Johnson actually drafted the proposed budget for Page's approval: "I am greatly in need of this revised form [budget] with your signature." See Johnson to Page, Mar. 25, 1939, in NA, RG 115, box 1098, File 301.1.

7. E. H. Heinemann, "Enlargement of Marshall Ford Dam," *The Reclamation Era* 28 (Dec., 1938): 341; John C. Page, "Appropriations, Fiscal Year 1940," *The Reclamation Era* 39 (Mar., 1939): i; "Sixteen Dams Being Built by

Reclamation Bureau," *Engineering News-Record* 121 (Dec., 1938): 746; Lower Colorado River Authority, Minutes of the Board of Directors, Oct. 7, 1940, 1507–23. Nearly 50 percent of the 1940 appropriation to the bureau went to California and Washington. Also, the official announcement of a "change order" in the contract to Brown and Root, Inc., and the McKenzie Construction Company was not made until August 28, 1939, by Secretary Ickes. See Press Release, Aug. 28, 1939, Reclamation Records, in NA, RG 115, box 1088, Press Release File 023.6.

8. Richard Lowitt, *The New Deal and the West*, 15–17; George Norris Green, *The Establishment in Texas Politics: The Primitive Years, 1938–1957*, 15. See also introduction by Don Graham in Work Projects Administration, *Texas: A Guide to the Lone Star State* (rpt., Austin, 1986), and Henry Morgenthau, Diaries, Jan., 1938, book 108, pp. 178, 183, and Feb., 1938, book 109, p. 75.

9. Filler-CRA Light, n.d., in Lyndon B. Johnson, House Papers, box 166, Colorado River District File; Curtis Cox interview with author, College Station, Texas, July 16, 1987. Much of the LCRA publicity was specifically directed: "Mother, the working woman at home: Statistics show that a farm woman carries approximately 50 tons of water a year. That means, for this chore alone, a full month of eight hour days and about 200,000 unnecessary steps in the process. An electric water pressure system can eliminate such waste at a power consumption rate of twenty to thirty kilowatt hours a month. At 2.5 cents a kilowatt hour, normal public power rates, this work can be done at fifty to seventy five cents a month. Pretty low wages for Mother, indeed."

10. Norris to Cooke, Oct. 24, 1935, in Franklin D. Roosevelt, Rural Electrification Administration Papers [hereafter, FDR, REA Papers], of 1570, box 1; J. P. Schaenzer, *Rural Electrification*, 1–11.

11. Filler-CRA Light, n.d., in Lyndon B. Johnson, House Papers, box 166, Colorado River District File. See also TVA Papers in Franklin D. Roosevelt Papers [hereafter, FDR, TVA Papers], boxes 1, 2, 3.

12. Jean Christie, "Morris L. Cooke and Energy for America," in Carroll W. Pursell, Jr., *Technology in America: A History of Individuals and Ideals*, 202, 212; Norris to Cooke, Oct. 24, 1935, in FDR, REA Papers, of 1570, box 1; Gifford Pinchot, *The Fight for Conservation*, 1–147; Ickes to Norris, Jan. 10, 1934, in George Norris Papers, box 63.

13. Jean Christie, *Morris Llewellyn Cooke: Progressive Engineer*; E. Taylor Parks and Lois F. Parks, eds., *Memorable Quotations of Franklin D. Roosevelt*, 108; D. Clayton Brown, *Electricity for Rural America: The Fight for the REA*, 22–25.

14. David Lilienthal, "Progress in the Electrification of the American Home and Farm," Sept. 19, 1934, in FDR, TVA Papers, box 2, TVA-1934 File.

15. Caro, *Years of Lyndon Johnson*, 502–15.

16. Cooke to Roosevelt, Aug. 2, 1935, in FDR, REA Papers, Official File 1570, box 1; Executive Order no. 7130, Prescribing Rules and Regulations Relating to Approved Projects Administered and Supervised by the Rural Electrification Administration under the Emergency Relief Act of 1935, The White

House, Aug. 7, 1935, ibid. REA Executive Order no. 7037 was signed on May 11, 1935. A comprehensive bill introduced by Senator George Norris and Representative Sam Rayburn was passed by Congress on May 20, 1936, and became the Rural Electrification Act of 1936.

17. *1938 Report of the Rural Electrification Administration* (Washington, D.C., 1939), 177–79 found in John W. Carmody Papers, box 93.

18. Ibid. See also FDR, REA Papers, boxes 1, 2, 3.

19. Carmody to Roosevelt, Feb. 4, 1939, in FDR, REA Papers, box 2; H. S. Person, "The Rural Electrification Administration in Perspective," *Agricultural History* 24 (Apr., 1950): 70–89; D. Clayton Brown, "Rural Electrification in the South, 1920–1955" (Ph.D. diss., University of California, Los Angeles, 1970), 140–97.

20. Merle Miller, *Lyndon: An Oral Biography*, 71; Alfred Steinberg, *Sam Johnson's Boy: A Close-up of the President from Texas*, 132.

21. Samuel I. Rosenman, ed., *The Public Papers and Address of Franklin D. Roosevelt*, vol. 1, 3; James MacGregor Burns, *Roosevelt: The Lion and the Fox*, 46, 98, 155; Nathan Miller, *FDR: An Intimate History*, 231–33.

22. State of New York, *Public Papers of Franklin D. Roosevelt, 1929*, 11–13; idem, *Public Papers of Franklin D. Roosevelt, 1930*, 436, 444, 448; Franklin D. Roosevelt, Statement of the Governor, Jan. 14, 1930, in Norris Papers, box 248; State of New York, *Message of the Governor Relative to Development of the State's Water Power Resources on the St. Lawrence*, 1.

23. State of New York, *Public Papers of Franklin D. Roosevelt, 1931*, 42–43, 80–85, 590–93; *New York Times*, Nov. 9, 1931; *New York Herald Tribune*, May 10, 1931.

24. *New York World Telegram*, May 13, 1931.

25. John Bauer, "If Power Cost Nothing to Produce—What Would Be the Cost to the User?" n.p., n.d., in Morris Llewellyn Cooke Papers, box 247, Distribution and Literature File; Morris Cooke, "Paying Too Much for Electricity," *The New Republic* (Dec. 21, 1932), report found in ibid.

26. "Roosevelt on the Shoal," n.p., 1929, found in Norris Papers, box 248; "U.S. Ownership of Shoals Urged by Roosevelt," n.p., 1929, ibid.

27. Arthur Schlesinger, *The Age of Roosevelt: The Coming of the New Deal*, 320–25; Preston J. Hubbard, *Origins of the TVA: The Muscle Shoals Controversy, 1920–1932*, 314–15. For a negative view on the evils of the TVA see Dean Russell, *The TVA Idea* (New York, 1949), 1–63.

28. Roosevelt to Norris, Dec. 14, 1932, in Franklin D. Roosevelt, Papers of the Governor of New York State, 1929–1932, box 60, Series 1 Norr-Nort. File; Frank Freidel, *Franklin D. Roosevelt: Launching the New Deal*, 350–54; Twentieth Century Fund, *Electric Power and Government Policy*, 573–678; Paul K. Conklin, *The New Deal*, 46–48; Richard Lowitt, "The TVA, 1933–1945" in Erwin C. Hargrave and Paul K. Conklin, eds., *TVA: Fifty Years of Grass Roots Bureaucracy*, 35–65.

29. Freidel, *Franklin D. Roosevelt*, 351–54. Freidel points out that the idea of public generation and distribution was not necessarily a radical idea and that it dated from the Progressive Era; it was Owen D. Young of General Electric who suggested to Roosevelt that "if consumers could not obtain sat-

isfactory rates, municipalities [and cooperatives] should establish their own power companies." When completed, the TVA would have twenty-one dams.

30. John R. Moore, ed., *The Economic Impact of TVA*, 10–12. Roosevelt had a number of detailed studies conducted to ensure the sound creation of the Tennessee River project. See Glavis to Ickes, Apr. 14, 1933, in FDR, TVA Papers, box 1, Glavis-Thompson File; and Huston Thompson, "Confidential Report to the President," n.p., n.d., ibid. See also Edgar B. Nixon, *Franklin D. Roosevelt and Conservation 1911–1945*, vol. 1, 151–52; Joseph S. Ransmeier, *The Tennessee Valley Authority: A Case Study in the Economics of Multiple Purpose Stream Planning*, 50–65; Raymond Moley, *The First New Deal*, 323–24.

31. McCraw, *TVA and the Power Fight*, 51–54. See also McSwain to President, Mar. 14, 1933, in FDR, TVA Papers, box 1, TVA-1933 File. It appears Roosevelt had a direct hand in selecting the name of the Tennessee River project. The original bill, HR 1672, referred to the new power producer as the "Tennessee Development Authority" or "Tennessee River Valley Authority."

32. McCraw, *TVA*, 59, 73; Twentieth Century Fund, *Electric Power*, 601–52. See John E. Babcock interview, Nov. 22, 1983.

33. Statement of Lyndon B. Johnson, Apr. 12, 1938, in LCRA Papers, box PB 26, General Information, U.S. Government, Lyndon B. Johnson, 1937–1938, File; *Fort Worth Star Telegram*, Mar. 6, 1941; Congressional Resolution, Feb. 24, 1941, in NA, RG 115, box 1088, Clippings File.

34. Interview with Congressman Lyndon B. Johnson, *CRA Light*, n.d., in Lyndon B. Johnson, House Papers, box 166, Colorado River District File.

35. Wirtz to Johnson, May 18, 1938, ibid., LCRA File; Texas Power and Light Company to Our Customers, Sept. 17, 1938, ibid.

36. Federal Emergency Administration of Public Works, Press Release, no. 3517, n.d., in NA, RG 115, box 1088, File 023.6.

37. Starcke to Johnson, Sept. 21, 1939, in Max Starcke Papers, box 10, File 1938; Johnson to Gideon, Sept. 21, 1939, ibid.; Caro, *Years of Lyndon Johnson*, 525–27; Lower Colorado River Authority, Minutes of the Board of Directors, May 27, 1940, 1371–97.

38. Babcock interview.

39. Radio Comments, Sept. 23, 1938, in Lyndon B. Johnson, House Papers, box 166, LCRA-Buchanan Dam File; Starcke to Press, Sept. 21, 1939, in Starcke Papers, box 10, File 1938; *Austin American*, Sept. 24, 1938. On September 27, 1938, a telegram arrived at the Pedernales Electric Cooperative (PEC) from the REA granting a loan of $1,322,000.

40. Johnson to Roosevelt, July 18, 1939, in Franklin D. Roosevelt Papers, Official Files [hereafter, FDR, Official Files], box 1, Colorado River, 1933–39, File 482a; Wirtz to Ickes, Oct. 30, 1939, in NA, RG 115, box 1088, Clippings File; Ickes to Wirtz, Nov. 15, 1939, in NA, RG 115, box 1088, Clippings File; *Dallas Morning News*, Sept. 2, 1939; *Wall Street Journal*, Sept. 7, 1939; Lower Colorado River Authority, Minutes of the Board of Directors, Dec. 23, 1940, 1880–1900; Gordon Fulcher interview, Jan. 13, 1969. The purchase price was $5 million raised through the issuance of bonds that the Reconstruction Finance Corporation agreed to purchase. The sixteen-county area included

San Saba, Lampasas, Llano, Burnet, Blanco, Travis, Hays, Guadalupe, Caldwell, Bastrop, Williamson, Kerr, Lee, Fayette, Washington, and Colorado counties.

41. An Austin newspaper commented that the LCRA will soon "establish" power-rate yardsticks that will influence power rates all over the United States (found in NA, RG 115, box 1088, Clippings File). See also Sim Gideon interview, Mar. 15, 1975.

42. Roosevelt to Johnson, July 20, 1939, in FDR, Official Files, box 1, File 482a.

43. Johnson to Roosevelt, July 29, 1939, in FDR, REA Papers, box 3, File 1939, F-2; Roosevelt to Johnson, Aug. 2, 1939, ibid.

44. Ickes to Roosevelt, Jan. 2, 1940, in FDR, Official Files, box 6, Interior Department; *The Reclamation Era* 30 (Apr., 1940): 108; Steinberg, *Sam Johnson's Boy*, 143–44.

45. Harold L. Ickes, *The Secret Diary of Harold L. Ickes*, vol. 3, *The Lowering Clouds 1939–1941*, 95. See also Memo, Jan. 29, 1974, W-28, in Mary Rather Reminiscences, box 1.

46. "Under Secretary Wirtz on Power," *The Reclamation Era* 30 (Mar., 1940): i. See also Pacific Constructors, Inc., *Shasta Dam and Its Builders*; Donald Worster, *Rivers of Empire: Water, Aridity, and the Growth of the American West*, 191–248.

47. Alvin J. Wirtz Address, Mar. 15, 1940, Portland, Oregon, in Alvin J. Wirtz Papers, box 8, Bonneville Speech on Power File.

48. Wirtz to Norris, Aug. 22, 1941, in George Norris Papers, box 253; Address of A. J. Wirtz of Austin, Texas, delivered at Roosevelt-for-President Rally, Dallas, Texas, Apr. 24, 1940, in LCRA Papers, box PB 25.

49. Wirtz to Norris, Aug. 22, 1941, in George Norris Papers, box 253. The term, "Texas' Little TVA," if not coined by Wirtz was certainly promoted by the always image-minded undersecretary. After his appointment to the Interior Department the term comparing LCRA with TVA appears in numerous documents. For example, see Ickes to President, Feb. 21, 1941, in Wirtz Papers, box 10, LCRA-CP&L File.

50. Fulcher interview, Jan. 13, 1969; Miller, *Lyndon*, 83–86; Ickes, *Diary*, vol. 3, 526; Provence, *Lyndon B. Johnson*, 43–47; Memo, Aug. 2, 1973, W-37, in Rather, Reminiscences, box 1. In addition to Wirtz, John B. Connally and Tom Miller were Johnson's key campaign workers in the Senate race. Ickes had encouraged Wirtz not to leave the Interior Department. In his diary, Ickes indicates that the president had directly instructed him "to tell Wirtz to take [only] a leave of absence" for whatever time was needed for the Johnson campaign. A year earlier, in May, 1940, when Roosevelt and Ickes discussed the possibility of his moving from secretary of the interior to secretary of war, the president rejected the idea according to Ickes: "I would like to move you . . . but I don't know where I could find a successor for Interior." Ickes' reply was: "You could safely make Wirtz Secretary of Interior." See Ickes, *Diary*, vol. 3, 186–87, 502, 526.

51. Lilienthal to Wirtz, Mar. 26, 1940, in Wirtz Papers, box 11, LCRA Miscellaneous File; Wirtz to Starcke, Feb. 27, 1940, ibid.

52. Press Release, P.N. 66975, June 24, 1939, in NA, RG 115, box 1088, File 023.6; Press Release, Jan. 24, 1941, ibid., box 1089, File 101.03; Press Release, P.N. 12168, Dec. 29, 1937, ibid., box 908, File 023.6; John Moulton interview with author, Knoxville, Tennessee, Aug. 2, 1987. Although larger in size, Marshall Ford is almost an identical twin of the Norris Dam built on the Clinch River as part of the TVA project. In 1939 Marshall Ford was fourth in size in the United States; in 1940 it ranked fifth behind Hoover, Shasta, Grand Coulee, and Friant. See "World's Fifth Largest Concrete Dam Completed," Nov. 22, 1941, in NA, RG 115, box 1088, File 023.6.

53. *The Reclamation Era* 30 (June, 1940): 181; General Information Concerning the Colorado River Project, Texas, Apr. 15, 1940, in NA, RG 115, box 1102, General Correspondence in Power Transmission Lines File; *Houston Chronicle*, Aug. 4, 1940. Marshall Ford (Mansfield Dam) was completed on June 1, 1941.

54. Rural Electrification Administration, *Report of the Administration of REA-1940* (Washington, D.C., 1941), 54, found in FDR, REA Papers, box 2.

55. Lower Colorado River Authority films, *A History of Progress*; *The Power of a River: The Story of Harnessing the Giant Colorado River of Texas*; *The LCRA: A Family Affair*; *Reflections of a River*. See also John Williams, *The Story of the Lower Colorado River Authority*, 8–11; Lower Colorado River Authority, *LCRA Fact Book* (May, 1987); idem, *LCRA Fact Book* (July, 1987); idem, *1985 Pocket Facts*; idem, *Highland Lakes and Dams*.

56. Lyndon B. Johnson Address, Oct. 24, 1957, Austin, Texas, in Lyndon B. Johnson Papers, LBJ-A, box 83, LCRA Dinner File.

Bibliography

For the convenience of the reader the bibliography has been divided into nine parts: manuscripts, archival collections, public documents, interviews, newspapers, films, books, articles, and court decisions. This list does not include all the government documents, pamphlets, articles, interviews, and books that were considered in the course of preparing this study, yet it was compiled with the hope that no pertinent material would be omitted. The most valuable sources were the Lower Colorado River Authority Papers, the Alvin J. Wirtz Papers, and the early congressional records of Lyndon B. Johnson, all housed at the Lyndon B. Johnson Presidential Library in Austin, Texas. Key background material on the Bureau of Reclamation, the Department of Interior, and the Public Works Administration is located at the National Archives. The Library of Congress houses the Woodrow Wilson Papers, Harold L. Ickes Papers, and George Norris Papers. The Franklin D. Roosevelt Presidential Library in Hyde Park, New York, and the Herbert C. Hoover Presidential Library in West Branch, Iowa, both contain material on all aspects of the depression, federal hydroelectric development, and the New Deal. A chronicle of the minutes of the Lower Colorado River Authority is located at its headquarters on Town Lake in Austin, with additional material available at the Texas State Archives, the Austin History Center, and the Eugene C. Barker Texas History Center.

MANUSCRIPT COLLECTIONS

Buchanan, James P. Papers. Eugene C. Barker Texas History Center, Austin, Texas.

Carmody, John M. Papers. Franklin D. Roosevelt Library, Hyde Park, New York.

Cooke, Morris Llewellyn. Papers. Franklin D. Roosevelt Library, Hyde Park, New York.

BIBLIOGRAPHY 143

Hoover, Herbert C. Papers. Herbert Hoover Library, West Branch, Iowa.
Ickes, Harold L. Papers. Manuscript Division, Library of Congress, Washington, D.C.
Johnson, Lyndon B. Papers. Lyndon B. Johnson Library, Austin, Texas.
Morgenthau, Henry. Diaries. Franklin D. Roosevelt Library, Hyde Park, New York.
Norris, George. Papers. Manuscript Division, Library of Congress, Washington, D.C.
Rather, Mary. Reminiscences. Lyndon B. Johnson Library, Austin, Texas.
Roosevelt, Franklin D. Papers. Franklin D. Roosevelt Library, Hyde Park, New York.
Starcke, Max. Papers. Austin History Center, Austin, Texas.
Wilson, Woodrow. Papers. Manuscript Division, Library of Congress, Washington, D.C.
Wirtz, Alvin J. Papers. Eugene C. Barker Texas History Center, Austin, Texas.

ARCHIVAL COLLECTIONS

Lower Colorado River Authority Papers. Lyndon B. Johnson Library, Austin, Texas.
National Archives. Record Group 69. Records of the Work Projects Administration, Washington, D.C.
National Archives. Record Group 77. Records of the Army Corps of Engineers, Fort Worth, Texas, Washington, D.C.
National Archives. Record Group 115. Records of the Bureau of Reclamation, Washington, D.C.
National Archives. Record Group 138. Records of the Federal Power Commission, Fort Worth, Texas, Washington, D.C.
Tennessee Valley Authority. Papers. Franklin D. Roosevelt Library, Hyde Park, New York.
Texas State Archives. Record Group 303. Records of State Purchasing and General Services Commission, Austin, Texas.

PUBLIC DOCUMENTS

Congressional Record. 57th Cong., 2d sess., 1903. Washington, D.C.
———. 60th Cong., 1st and 2d sess., 1908–1909. Washington, D.C.
———. 63rd Cong., 1st, 2d, and 3d sess., 1913–1915. Washington, D.C.
———. 65th Cong., 3d sess., 1919. Washington, D.C.
———. 66th Cong., 1st and 2d sess., 1919–1920. Washington, D.C.
———. 67th Cong., 1st and 2d sess., 1921–1922. Washington, D.C.
———. 75th Cong., 1st sess., 1937. Washington, D.C.
Lower Colorado River Authority. *Highland Lakes and Dams.* Austin, 1983.
———. *LCRA Fact Book.* Austin, May, 1987.
———. *LCRA Fact Book.* Austin, July, 1987.
———. *Lower Colorado River Authority: 1985 Annual Report.* Austin, 1986.
———. *Lower Colorado River Authority: 1986 Annual Report.* Austin, 1987.
———. *Minutes of the Board of Directors.* Austin, 1935–1942.

———. *1985 Pocket Facts.* Austin, 1985.
———. *The Lower Colorado River Authority.* Austin, 1949.
———. *What Is the Lower Colorado River Authority?* Austin, ca. 1941.
Proceedings of a Conference of Governors in the White House Washington, D.C., May 13–15, 1908. Washington, D.C., 1909.
Proceedings of Second National Conservation Congress St. Paul, September 5–8, 1910. Saint Paul, 1910.
Special Message of the President of the United States, Transmitted to the Two Houses of Congress. 61st Cong., 2d sess., no. 533, 1910.
State of New York. *Message of the Governor Relative to Development of the State's Water Power Resources on the St. Lawrence.* Albany, 1929.
———. *Public Papers of Franklin D. Roosevelt, 1929.* Albany, 1929.
———. *Public Papers of Franklin D. Roosevelt, 1930.* Albany, 1931.
———. *Public Papers of Franklin D. Roosevelt, 1931.* Albany, 1937.
State of Texas. *General Laws.* 38th Leg., 2d C.S., 1923.
———. *House Journal.* 43d Leg., 3d sess., 1933–1934.
———. *Senate Journal.* 43d Leg., 1st, 2d, and 3d sess., 1933–1934.
———. *Constitution.* 1876.
U. S. Congress. House of Representatives. *Flood Control on the Colorado River, Texas.* 66th Cong., 1st sess., 1919, H. Doc. 304.
———. *Hearings on War Expenditures.* House Select Committee on Expenditures in the War Department. 66th Cong., 2d sess., 1920.
———. *Preliminary Examination of Colorado River, Texas with a View to Devising Plans for Flood Protection.* H. Doc. 304. 66th Cong., 1st sess., 1919.
———. *Preliminary Examination of Colorado River, Texas, from Its Mouth As Far As Is Practicable, with a View to Removing the Raft.* 66th Cong., 2d sess., 1919, H. Doc. 529.
———. *Report of the Examination of the Colorado River, Texas.* 51st Cong., 2d sess., 1891, H. Doc. 138.
———. *Report of the National Conservation Commission.* 60th Cong., 2d sess., 1909.
———. *Report of the Preliminary Examination of Colorado River, Texas.* 71st Cong., 2d sess., 1930, H. Doc. 361.
———. *Report of Examination of Guadalupe River, Texas.* 58th Cong., 2d sess., 1903, H. Doc. 187.
———. *Report of Examination of Neches River, Texas.* 60th Cong., 1st sess., 1908, H. Doc. 870.
———. *Report of Examination of Sabine River from Its Mouth to Logan's Port, Louisiana and Brazoria, Texas.* 60th Cong., 1st sess., 1908, H. Doc. 490.
———. *Report of Examination of Sulphur River, Texas.* 58th Cong., 2d sess., 1903, H. Doc. 231.
———. *Report of the Preliminary Examination of the Trinity River, Texas.* 51st Cong., 2d sess., 1891, H. Doc. 275.
———. *Report of Examination of Trinity River, Texas.* 58th Cong., 2d sess., 1903, H. Doc. 118.

———. *War Expenditures, Ordinance.* 66th Cong., 2d sess., 1920, H. Doc. 998.
———. *The Water Powers of Texas*, by T. U. Taylor, 58th Cong., 2d sess., 1904, H. Doc. 759.
U. S. Congress. Senate. *Annual Report of the Secretary of War.* 33d Cong., 1st sess., 1853, S. Doc. 1, vol. 2.
———. *Annual Report of the Secretary of War.* 33d Cong., 2d sess., 1854, Exec. Doc. 1, vol. 2.
———. *Preliminary Report of the Inland Waterways Commission.* 60th Cong., 1st sess., 1908.
U.S. Department of Agriculture. *Middle Colorado River Watershed, Texas.* 78th Cong., 1st sess., Washington, D.C., 1943, H. Doc. 270.
U.S. Department of Army, Chief of Engineers. *Annual Report.* Various years to include 1882, 1891, 1895, 1900, 1930, and 1940. Washington, D.C.
———. *Preliminary Examination of Matagorda Bay, Texas*, 59th Cong., 1st sess., 1905, H. Doc. 154.
———. *Preliminary Examination and Survey of Colorado River, Texas.* 60th Cong., 2d sess., 1900, H. Doc. 1211.
———. *Preliminary Examination of Colorado River, Texas, with a View to Its Improvements by Means of Locks and Dams.* 43rd Cong., 2d sess., 1875, H. Doc. 657.
———. *Preliminary Examination Colorado River, Texas, Having in View the Extent of Any Improvement of Said River with Reference to Snagging and Cleaning the River.* House Rivers and Harbors Committee, 63d Cong., 1st sess., 1913, H. Doc. 3.
———. *Preliminary Examination and Survey Colorado River, Texas, from Its Mouth As Far As Practicable, with a View to Removing the Raft.* 66th Cong., 2d sess., 1919, H. Doc. 529.
U.S. Geological Survey. *Floods of Central Texas in September 1921*, by C. E. Ellsworth, Water-Supply and Irrigation Papers, No. 488. Washington, D.C., 1923.
———. *Floods in the United States—Magnitude and Frequency.* No. 771. Washington, D.C. 1936.
———. *Floods in Texas of 1940.* Washington, D.C., 1941.
———. *Major Floods in Texas in 1935.* No. 796G. Washington, D.C. 1937.
———. *Summary of Surface Waters of Texas, 1898–1937.* No. 850. Washington, D.C., 1939.
———. *The Austin Dam*, by Thomas U. Taylor, Water-Supply and Irrigation Papers, No. 40. Washington, D.C., 1900.
———. *Water Supply and Irrigation Papers.* No. 28. Washington, D.C., 1899.
United States Statutes at Large. Washington, D.C., 1916–40.
U.S. War Department. *Report on the Fixation and Utilization of Nitrogen.* No. 2041. Washington, D.C., 1922.
———. *War Expenditures, Ordnance.* 66th Cong., 2d sess., 1920, H. Doc. 998.
Work Projects Administration. *Texas: A Guide to the Lone Star State.* Austin, 1986.

INTERVIEWS

Babcock, John E. Interview, November 22, 1983. LBJ Oral History Collection, Austin, Texas.
Bostic, Bobby. Interview with author. Buchanan Dam, Texas, Apr. 20, 1987.
Brown, George. Interviews, Apr. 6, 1968, Aug. 9, 1969, July 11, 1977. LBJ Oral History Collection, Austin, Texas.
Chapman, Oscar. Interview, Feb. 5, 1981. LBJ Oral History Collection, Austin, Texas.
Clayton, Jim. Interview with author. Austin, Texas, July 30, 1987.
Cox, Curtis. Interview with author. College Station, Texas, July 16, 1987.
Fulcher, Gordon. Interview, Jan. 13, 1969. LBJ Oral History Collection, Austin, Texas.
Gideon, Sim. Interview, Mar. 21, 1968. LBJ Oral History Collection, Austin, Texas.
Gloyna, Emmett. Interview with author. Austin, Texas, June 13, 1986.
Griffith, Llewellyn B., Sr. Interview, Aug. 15, 1978. LBJ Oral History Collection, Austin, Texas.
Hopkins, Welly K. Interviews, May 11, 1965, Nov. 14, 1968, June 9, 1977. LBJ Oral History Collection, Austin, Texas.
Moulton, John. Interview with author. Knoxville, Tennessee, Aug. 2, 1987.
Williams, John. Interviews with author. Austin, Texas, May 21, July 30, 1987.

NEWSPAPERS AND PERIODICALS

Arizona Republic, 1938.
Austin American, 1931–1951.
Austin American-Statesman, 1934–1936.
Austin Review, 1936.
Austin Semi-Weekly Statesman, 1900.
Austin Statesman, 1895–1940.
Austin Tri-Weekly Texas State Gazette, 1869.
Calexico Chronicle, 1937.
Chicago Daily Tribune, 1934.
Chicago Herald, 1926.
Compressed Air Magazine, 1933–1940.
Dallas Morning News, 1934–1951.
El Paso Times, 1937.
Engineering News-Record, 1800–1939.
Fort Worth Star Telegram, 1941.
Houston Chronicle, 1936–1940.
Houston Post Sentinel, 1936.
Los Angeles Times, 1938.
New York Herald Tribune, 1931.
New York Times, 1938–1939.
New York World Telegram, 1931.
Reclamation Era, 1931–1939.

San Angelo Evening Standard, 1931–1937.
San Angelo Standard Times, 1934.
San Antonio Express, 1931–1937.
San Antonio News, 1937.
Wall Street Journal, 1938–1939.
Washington Herald, 1936.
Washington Post, 1937–1938.
Washington Star, 1936.

DISSERTATIONS

Brown, D. Clayton. "Rural Electrification in the South 1920–1955." Ph.D. diss., University of California, Los Angeles, 1970.
Clay, Comer. "The Lower Colorado River Authority: A Study in Politics and Public Administration." Ph.D. diss., University of Texas, 1948.
Hill, G. C. "The History and Purpose of the Lower Colorado River Authority." Ph.D. diss., University of Texas, 1935.

FILMS

Lower Colorado River Authority. *A History of Progress*, Austin, Texas, 1985.
———. *The LCRA: A Family Affair*. Austin, Texas, n.d.
———. *The Power of a River: The Story of Harnessing the Giant Colorado River of Texas*. Austin, Texas, n.d.
———. *Reflections of a River*. Austin, Texas, n.d.

BOOKS

Abrams, Ernest R. *Power in Transition*. New York, 1940.
Baker, Eugene C. *The Life of Stephen F. Austin: Founder of Texas, 1793–1836*. Austin, 1949.
Barrows, H. K. *Floods: Their Hydrology and Control*. New York, 1948.
Baruch, Bernard M. *Baruch: The Public Years*. New York, 1960.
Blum, John Morton. *From the Morgenthau Diaries*. 3 vols. Boston, 1959.
———. *The Republican Roosevelt*. New York, 1972.
Bonbright, James C. *Public Power Policies*. New York, 1940.
Brown, D. Clayton. *Electricity for Rural America: The Fight for the REA*. Westport, Connecticut, 1980.
Brown, Frank. *Annals of Travis County and of the City of Austin: From the Earliest Times to the Close of 1875*. n.p., n.d.
Bunger, H. P. *Irrigation and Flood Protection, Austin to Matagorda, Texas No. 20*. Washington, D.C., n.d.
Burner, David. *Herbert Hoover: A Public Life*. New York, 1979.
Burns, James MacGregor. *Roosevelt: The Lion and the Fox*. New York, 1956.
Caro, Robert A. *The Years of Lyndon Johnson: The Path to Power*. New York, 1981.
Christie, Jean. "Morris L. Cooke and Energy for America." In Carroll W. Pursell, Jr., *Technology in America: A History of Individuals and Ideas*. Cambridge, 1982.

———. *Morris Llewellyn Cooke: Progressive Engineer*. New York, 1983.
Colorado River Improvement Association. *Statement of Committee*. Austin, 1915.
Conklin, Paul K. *The New Deal*. Arlington Heights, Illinois, 1975.
Cooke, Morris L. *What Price Electricity for Our Homes*. Philadelphia, 1928.
Daniel, Pete. *Deep'n as It Come: The 1927 Mississippi River Flood*. New York, 1972.
Dowell, C. L., and S. D. Breeding. *Dams and Reservoirs in Texas*. Austin, 1967.
Droze, Wilmon H., George Wolfskill, and William E. Leuchtenburg. *Essays on the New Deal*. Austin, 1969.
Fausold, Martin L. *The Presidency of Herbert C. Hoover*. Lawrence, Kansas, 1985.
Frank, Arthur DeWitt. *The Development of the Federal Program of Flood Control on the Mississippi River*. New York, 1930.
Freidel, Frank. *Franklin D. Roosevelt: Launching the New Deal*. Boston, 1973.
Golze, Alfred R. *Reclamation in the United States*. Caldwell, Idaho, 1952.
Green, George Norris. *The Establishment in Texas Politics: The Primitive Years, 1938–1957*. Westport, Connecticut, 1979.
Haley, J. Evetts. *A Texan Looks at Lyndon: A Study in Illegitimate Power*. Canyon, Texas, 1964.
Harbaugh, William Henry. *Power and Responsibility: The Life and Times of Theodore Roosevelt*. New York, 1975.
Hardeman, D. B., and Donald C. Bacon. *Rayburn: A Biography*. New York, 1987.
Haw, John W., and F. E. Schmitt. *Report on Federal Reclamation to the Secretary of the Interior*. Washington, D.C., 1934.
Hawley, Ellis. *Herbert Hoover as Secretary of Commerce*. Iowa City, 1981.
Hays, Samuel P. *The Response to Industrialism, 1885–1914*. Chicago, 1957.
———. *Conservation and the Gospel of Efficiency: The Progressive Conservation Movement 1890–1920*. Cambridge, 1959.
Hendrickson, Kenneth E. *The Waters of the Brazos: History of the Brazos River Authority, 1929–1979*. Waco, Texas, 1981.
Hollon, W. Eugene. *The Southwest: Old and New*. New York, 1961.
Holt, W. Stull. *The Office of Chief Engineers of the Army*. Baltimore, 1923.
Hoover, Herbert C. *The Memoirs of Herbert Hoover: The Great Depression, 1929–1941*. New York, 1952.
Horgan, Paul. *Great River: The Rio Grande in North American History*. Austin, 1984.
Hoyt, William G., and Walter B. Langbein, *Floods*. Princeton, 1955.
Hubbard, Preston J. *Origins of the TVA: The Muscle Shoals Controversy, 1920–1932*. Nashville, 1961.
Hundley, Norris, Jr. *Water and the West: The Colorado River Compact and the Politics of Water in the American West*. Berkeley, 1975.
Ickes, Harold L. *Back to Work*. New York, 1935.
———. *The Autobiography of a Curmudgeon*. New York, 1943.
———. *The Secret Diary of Harold L. Ickes*. Vol. 3, *The Lowering Clouds, 1939–1941*. New York, 1953.

Isakoff, Jack F. *The Public Works Administration*. Urbana, Illinois, 1938.
Jarrett, Henry, ed. *Perspectives on Conservation*. Baltimore, 1966.
Johnson, Robert Underwood. *Remember Yesterdays*. Boston, 1923.
Keller, Charles Colonel. *Crops of Engineers: The Power Situation during the War*. Washington, D.C., 1921.
Kerwin, Jerome G. *Federal Water Power Legislation*. New York, 1926.
King, Judson. *The Conservation Fight: From Theodore Roosevelt to the Tennessee Valley Authority*. Washington, D.C., 1959.
Lamm, Richard D., and Michael McCarthy. *The Angry West: A Vulnerable Land and Its Future*. Boston, 1982.
Larson, Everett H., and David L. Goodman. "Reclamation Engineering." In *Dams and Control Works*, Washington, D.C., 1954.
Lavender, David. *Colorado River Country*. New York, 1982.
Leuchtenburg, William E. *Franklin D. Roosevelt and The New Deal, 1932–1940*. New York, 1963.
Lilienthal, David E. *TVA: Democracy on the March*. New York, 1953.
Long, Walter E. *Flood to Faucet*. N.p, 1956.
Lowitt, Richard. *George W. Norris: The Persistence of a Progressive, 1913–1933*. Urbana, Illinois, 1971.
———. "The TVA, 1933–1945." In Erwin C. Hargrave and Paul K. Conklin, eds., *TVA: Fifty Years of Grass Roots Bureaucracy*. Urbana, Illinois, 1983.
Lowry, R. L. *Flood Control by Marshall Ford Reservoir-Colorado River Project, Texas*. Washington, D.C., 1937.
Maass, Arthur. *Muddy Waters: The Army Engineers and the National Rivers*. Cambridge, 1951.
Maass, Arthur, and Raymond L. Anderson. *. . . and the Desert Shall Rejoice: Conflict, Growth and Justice in the Arid Environments*. Cambridge, 1978.
McCall, Edith. *Conquering the Rivers: Henry Miller Shreve and the Navigation of America's Inland Waterways*. Baton Rouge, 1984.
McCraw, Thomas K. *TVA and the Power Fight*. New York, 1971.
McCullough, David. *Mornings on Horseback*. New York, 1981.
McDonald, Forest. *Insull*. Chicago, 1962.
McElvaine, Robert S. *The Great Depression in America: 1929–1941*. New York, 1961.
McJimsey, George. *Harry Hopkins: Ally of the Poor and Defender of Democracy*. Cambridge, 1987.
McKay, Seth S., and Odie B. Faulk. *Texas after Spindletop*. Austin, 1965.
Major, John. *The New Deal*. New York, 1967.
Mead, Daniel W. *Report on the Dam and Water Power Development at Austin, Texas*. Madison, 1917.
Miller, Merle. *Lyndon: An Oral Biography*. New York, 1980.
Miller, Nathan. *FDR: An Intimate History*. New York, 1983.
Moeller, Beverly B. *Phil Swing and Boulder Dam*. Berkeley, 1971.
Moley, Raymond. *The First New Deal*. New York, 1966.
Moore, John R. ed. *The Economic Impact of TVA*. Knoxville, 1967.

Morgan, Arthur E. *Dams and Other Disasters: A Century of the Army Corps of Engineers in Civil Works*. Boston, 1971.
Morris, Edmond. *The Rise of Theodore Roosevelt*. New York, 1979.
Mowry, George E. *The Era of Theodore Roosevelt and the Birth of Modern America, 1900–1912*. New York, 1958.
Nash, Roderick. *Wilderness and the American Mind*. New Haven, 1967.
Nixon, Edgar B. *Franklin D. Roosevelt and Conservation 1911–1945*. 2 vols. Hyde Park, 1957.
Norris, George. *Fighting Liberal: The Autobiography of George W. Norris*. New York, 1946.
Pacific Constructors, Inc. *Shasta Dam and Its Builders*. San Francisco, 1945.
Parks, E. Taylor, and Lois F. Parks, eds. *Memorable Quotations of Franklin D. Roosevelt*. New York, 1965.
Patenaude, Lionel V. *Texans, Politics, and the New Deal*. New York, 1983.
Patterson, James T. *Congressional Conservatism and the New Deal: The Growth of the Conservative Coalition in Congress, 1933–1939*. Louisville, 1967.
———. *The New Deal and the States: Federalism in Transition*. Princeton, 1969.
Perkins, Dexter. *The New Age of Franklin Roosevelt, 1932–1945*. Chicago, 1957.
Pfeffer, E. Louise. *The Closing of the Public Domain*. Stanford, 1951.
Pinchot, Gifford. *The Fight for Conservation*. New York, 1910.
Provence, Harry. *Lyndon B. Johnson*. New York, 1965.
Ragsdale, Kenneth B. *The Year America Discovered Texas: Centennial '36*. College Station, 1987.
Ransmeier, Joseph S. *The Tennessee Valley Authority: A Case Study in the Economics of Multiple Purpose Stream Planning*. Nashville, 1942.
Reisner, Marc. *Cadillac Desert: The American West and Its Disappearing Water*. New York, 1986.
Richardson, Elmo R. *The Politics of Conservation: Crusades and Controversies, 1897–1913*. Berkeley, 1962.
Romasco, Albert U. *The Politics of Recovery*. New York, 1983.
———. *The Poverty of Abundance*. New York, 1965.
Roosevelt, Theodore. *Autobiography*. New York, 1913.
———. *Outdoor Pastimes of an American Hunter*. New York, 1926.
———. *Ranch Life and the Hunting Trail*. New York, 1888.
Rosenman, Samuel I. ed. *The Public Papers and Addresses of Franklin D. Roosevelt*. 9 vols. New York, 1938–1945.
Russell, Dean. *The TVA Idea*. New York, 1949.
Schaenzer, J. P. *Rural Electrification*. Milwaukee, 1955.
Schlesinger, Arthur. *The Age of Roosevelt: The Coming of the New Deal*. Boston, 1958.
Schrepfer, Susan R. *The Fight to Save the Redwoods*. Madison, 1983.
Steinberg, Alfred. *Sam Johnson's Boy: A Close-up of the President from Texas*. New York, 1968.

The State of the Union Messages of the Presidents: 1790–1966. New York, 1967.
Stiles, Henry R., ed., *Joutel's Journal of La Salle's Last Voyage, 1684–1687.* Albany, 1906.
Swain, Donald C. *Federal Conservation Policy, 1927–1933.* Berkeley, 1959.
Texas Almanac, 1980–1981. Dallas, 1979.
Thomas, B. F., and D. A. Watt. *The Improvement of Rivers: A Treatise on the Methods Employed for Improving Streams for Open Navigation, and for Navigation by Means of Locks and Dams.* New York, 1913.
Twentieth Century Fund. *Electric Power and Government Planning.* New York, 1948.
Webb, Walter Prescott. *The Great Plains.* New York, 1931.
———. *More Water for Texas: The Problem and the Plan.* Austin, 1954.
West, William B. *America's Greatest Dam.* New York, 1925.
Whisenhunt, Donald W. *The Depression in Texas: The Hoover Years.* New York, 1983.
White, Gilbert I. *Human Adjustment to Floods.* Chicago, 1945.
White, Graham, and John Maze. *Harold Ickes of the New Deal: His Private Life and Public Career.* Cambridge, 1985.
Williams, John. *The Story of the Lower Colorado River Authority.* Austin, 1985.
Williams, Kerwin J. *Grants-in-Aid under the Public Works Administration.* New York, 1939.
Wilson, Joan H. *Herbert Hoover: Forgotten Progressive.* Boston, 1975.
Worster, Donald. *Rivers of Empire: Water, Aridity and the Growth of the American West.* New York, 1985.

ARTICLES

Banks, Stanley C. "The Mormon Migration into Texas." *Southwestern Historical Quarterly* 49 (October, 1945): 233–44.
Barker, Eugene C. "Description of Texas by Stephen F. Austin." *Southwestern Historical Quarterly* 29 (October, 1924): 98–121.
Bates, Leonard J. "Fulfilling American Democracy: The Conservation Movement, 1907–1921." *Journal of American History* 44 (June, 1957): 29–57.
Bauer, John. "If Power Cost Nothing to Produce—What Would Be the Cost to the User?" n.p., n.d. In Franklin D. Roosevelt Papers. Franklin D. Roosevelt Library, Hyde Park, New York.
Brown, Rome G. "The Conservation of Water Powers." *Harvard Law Review* 26 (May, 1913): 601–30.
Chittenden, H. M. "Detention Reservoirs with Spillway Outlets as an Agency in Flood Control." *Proceedings of the American Society of Civil Engineers* (September, 1917): 1–23.
Christain, A. K. "Mirabeau Buonaparte Lamar." *Southwestern Historical Quarterly* 23 (April, 1920): 264–70.
Clay, Comer. "The Colorado River Raft." *Southwestern Historical Quarterly* 52 (April, 1949): 410–26.

Clements, Kendrick A. "Committee Disapproves Reservoir Level Law." *Engineering News-Record* 122 (March, 1939): 257.
———. "Herbert Hoover and Conservation, 1921–33." *American Historical Review* 89 (February, 1984): 67–74.
Cole, E. W. "La Salle in Texas." *Southwestern Historical Quarterly* 49 (April, 1946): 484–85.
"Colorado River Project, Texas." *The Reclamation Era* 26 (March, 1936): 71.
"Colorado River Project, Texas." *The Reclamation Era* 26 (May, 1936): 117.
"Colorado River Project, Texas." *The Reclamation Era* 26 (June, 1936): 145.
"Colorado River Project, Texas." *The Reclamation Era* 26 (August, 1936): 186.
"Committee Disapproves Reservoir Level Law." *Engineer News-Record* 122 (March 2, 1939): 257.
Cooke, Morris L. "Quaint Electrical Rates." *National Municipal Review* 19 (November, 1930): 3–7.
———. "Planning for Power." *The Nation* 134 (June, 1932): 621–24.
Crowther, Samuel. "Henry Ford Tackles a New Job." *Collier's* 74 (October 18, 1924): 5–6.
"Dedication of Marshall Ford Dam, Colorado River Project, Texas." *The Reclamation Era* 27 (April, 1937): 69–71.
Dent, P. W. "Texas Legislature in 1931." *The Reclamation Era* 23 (February, 1932): 24–26.
———. "Paying Too Much for Electricity." *The New Republic* 73 (December, 1932): 150–52.
Dodds, Gordon B. "The Historiography of American Conservation: Past and Prospects." *Pacific Northwest Quarterly* 56 (April, 1965): 75–81.
"Era Allotments to Bureau Reduced by $20,000,000." *The Reclamation Era* 25 (December, 1935): 232–33.
Fleming, Donald. "Roots of the New Conservation Movement." *Perspectives in American History* 6 (1972): 5–91.
"Full Bucket." *Time* 37 (August 8, 1938): 1–3.
"Funds for Texas Project Restored." *The Reclamation Era* 26 (February, 1936): 42.
Gressley, Gene M. "Arthur Powell Davis, Reclamation and the West." *Agricultural History* 43 (1968): 241–57.
Haimbaugh, George D., Jr. "The TVA Cases: A Quarter Century Later." *Indiana Law Journal* 41 (Winter, 1966): 197–227.
Hamilton, David E. "Herbert Hoover and the Great Drought of 1930." *Journal of American History* 68 (March, 1982): 850–75.
Hawley, Ellis. "Herbert Hoover, the Commerce Secretariat, and the Vision of an 'Associative State,' 1921–1929." *Journal of American History* 61 (1974–75): 117–140.
Heinemann, E. H. "Enlargement of Marshall Ford Dam." *The Reclamation Era* 28 (December, 1938): 341.
Hoover, Herbert. "The National Policy of the Development of Water Resources." *Port and Terminal* 6 (September, 1926): 9–14.
"Importance of Water Conservation." *The Reclamation Era* 26 (September, 1936): 20.

"It Will Probably Go Higher." *Engineering News-Record* 122 (March, 1939).
Keener, Kenneth B. "The Low Dam at Marshall Ford." *Engineering News-Record* 121 (December 1, 1938): 697–99.
"LCRA Announces Flood Policy." *Engineering News-Record* 122 (March 23, 1939): 406, 416.
LeDuc, Thomas. "The Historiography of Conservation." *Forest History* 9 (October, 1965): 23–28.
"Lower Colorado River Project, Texas." *The Reclamation Era* 25 (September, 1935): 187–90.
Lowitt, Richard. "A Neglected Aspect of the Progressive Movement: George W. Norris and Public Control of Hydroelectric Power, 1913–1919." *Historian* 27 (1965): 350–65.
"Marshall Ford Dam." *Construction Methods and Equipment* 19 (June, 1937): 64–65.
Mason, J. Rupert. "A New Deal for Reclamation." *The Reclamation Era* 25 (February, 1935): 25–26.
Morrison, G. W. "The Buchanan Dam." *Compressed Air Magazine* 41 (November, 1936): 5157–63.
"Multi-Purpose Dams Opposed in Texas." *Engineering News-Record* 122 (February 9, 1936): 6.
"A New Dam Goes into Service." *Engineering News-Record* 119 (July, 1937): 39.
"The New Deal in Review, 1936–1940." *The New Republic* 47 (May, 1940): 705–708.
Neyland, R. R. "A Few Facts about the Tennessee River and the Wilson Dam." n.p., n.d. In Franklin D. Roosevelt. Papers. Franklin D. Roosevelt Library, Hyde Park, New York.
Noggle, Burl. "The Origins of the Teapot Dome Investigation." *Journal of American History* 44 (September, 1957): 237–50.
"One Hundred Million Dollars Set Aside for Reclamation." *The Reclamation Era* 25 (August, 1935): 153.
Orum, Anthony M. "Enter Lyndon Johnson." *The Texas Observer* 77 (January 11, 1985): 13–19.
Page, John C. "Appropriations, Fiscal Year 1940." *The Reclamation Era* 28 (March, 1939): i.
———. "The Multiple-Purpose Project." *The Reclamation Era* 29 (May, 1939): 93–95.
Penick, James L., Jr. "Louis Russell Glavis: A Postscript to the Ballinger-Pinchot Controversy." *Pacific Northwest Quarterly* 55 (April, 1964): 67–75.
Person, H. S. "The Rural Electrification Administration in Perspective." *Agriculture History* 24 (April, 1950): 70–89.
Pinchot, Gifford. "The Long Struggle for Effective Federal Water Power Legislation." *The George Washington Law Review* 14 (December, 1945): 9–20.
Pisani, Donald J. "Forest and Conservation, 1865–1890." *Journal of American History* 72 (September, 1985): 340–59.

———. "Enterprise and Equity: A Critique of Western Water Law in the Nineteenth Century." *Western Historical Quarterly* 27 (January, 1987): 15–37.
Raines, C. W. "Enduring Laws of the Republic of Texas." *The Texas State Historical Association Quarterly* 2 (October, 1898): 155–58.
Reeves, William D. "PWA and Competitive Administration in the New Deal." *Pacific Historical Review* 24 (November, 1965): 457.
Roberts, O. M. "The Capitols of Texas." *The Texas State Historical Association Quarterly* 2 (October, 1898): 117–19.
Roose, Kenneth D. "The Recession of 1937–38." *The Journal of Political Economy* 56 (June, 1948): 239–48.
Sanford, George O. "Dams—High, Large, and Unusual, Part 1." *The Reclamation Era* 23 (February, 1932): 28–30.
———. "Dams—High, Large, and Unusual, Part 2." *The Reclamation Era* 23 (March, 1932): 58–60.
Sautter, Udo. "Government and Unemployment: The Use of Public Works before the New Deal." *Journal of American History* 73 (June, 1986): 59–81.
"Sixteen Dams Being Built by Reclamation Bureau." *Engineering News-Record* 121 (December, 1938): 746.
Slavick, Walter K. M. "Monuments to the Living." *The Reclamation Era* 30 (February, 1940): 42–45.
Swain, Donald C. "The Bureau of Reclamation and the New Deal, 1933–1940." *Pacific Northwest Quarterly* 61 (July, 1970): 137–46.
Swidler, Joseph C., and Robert H. Marquis. "TVA in Court: A Study of TVA's Constitutional Litigation." *Iowa Law Review* 32 (January, 1947): 296–326.
Terrell, Alex W. "The City of Austin from 1839 to 1865." *The Texas Historical Association Quarterly* 14 (October, 1910): 113–28.
Thompson, Nan T. "The Muddy Brazos in Early Texas." *Southwestern Historical Review* 63 (October, 1959): 239.
"Three Years of P.W.A." *The Reclamation Era* 27 (March, 1937): i.
"Towers Used on Topographic Survey of the Marshall Ford Reservoir Site, Colorado River Project, Texas." *The Reclamation Era* 27 (May, 1937): 108–109, 118.
"Undersecretary Wirtz on Power." *The Reclamation Era* 30 (March, 1940): 1.
Wilson, Richard Guy. "The Machine-Age Iconography in the American West: The Design of Hoover Dam." *Pacific Historical Review* 54 (November, 1985): 468–93.
Winkler, Ernest W. "The Permanent Location of the Seat of Government." *The Texas State Historical Association Quarterly* 10 (January, 1907): 207–25.

COURT CASES

Alabama Power Co. v. Ickes et al., 302 U.S. 464 (1938).
Ashwander et al. v. Tennessee Valley Authority et al., 297 U.S. 288 (1936).
Duke Power Co. et al. v. Greenwood County et al., 302 U.S. 485 (1938).
Escanaba & L. M. Transport Co. v. Chicago, 107 U.S. 678, 682 (1883).

Gibson v. U.S., 166 U.S. 269 (1897).
Gilman v. Philadelphia, 3 Wall, 713, 725 (1866).
LCRA v. Gulf Coast Water Co., 107 S.W. 2d 1101 (1937).
LCRA v. McCraw, 83 S.W. 2d 629 (1935).
Motl v. Boyd, 286 S.W. 458 (1926).
TVA v. Tennessee Electric Power Company 306 U.S. 118, 139 (1939).

Index

References to illustrations are indicated by page numbers in italic type.

Agricultural Credit Corporation, 30
Alabama Power Company, 53
Alamo, funds for maintenance of, 29
Alexander, Charles H., Sr., 9, 13–14, 17
Allred, James V., 72
Alvin J. Wirtz Dam, 105
American Public Utilities Bureau, 98
American Red Cross, 88
Anderson, W. E., 34
Arnold Dam, 49, 59. *See also* Inks Dam
Associated General Contractors of America, 69
Austin, Stephen F., 5
Austin, Texas, 4, 14, 42; and Austin Dam, 76–79; and dam lease with LCRA, 77–78; early flooding in, 8–10; electrical power for, 77–78, 96; first . dam at, 9–10; and flood of 1938, 86–91; and flood of 1936, 61; selection of, as capital, 5–6; water supply for, 9, 11–12
Austin Chamber of Commerce, 36
Austin City Council, 74–75, 77–78
Austin Dam, 9–11, 12, 45, 49, 69, 74–75, 76, 78, 80–81, 86, 132n37
Austin Statesman, 69, 85

Balcones Escarpment, 4
Ballinger, Texas, 4
Bashore, Harry W., 90
Bastrop, Texas, 4, 25
Bauer, John, 98
Bluffton, Texas, 5
Bluffton Ferry, *11*
Bonneville Dam, 105

Bostic, Bobby, 127n61
Boulder Canyon Project Act, 26
Boulder Dam, xvii, 15–16. *See also* Hoover Dam
Brassos River, 5–6. *See also* Brazos River
Brazos River, 3, 5; location of state capital along, 5–6, 72
Brazos River Conservation and Reclamation District, 25, 34, 52
Brown, George, 67
Brown, Herman, 67, 68
Brown and Root, Inc., 67, 68, 73, 81, 87–88, 95
Brownwood, Texas, 8
Buchanan, James Paul ("Buck"), 25, 33; death of, 66, 110; and Elwood Mead, 49–50; and funding for LCRA, 34–36; and John Page, 60–61; and LCRA Bill, 37–39, 41; and Marshall Ford Dam funding, 59–60, 66–67; versus private utilities, 56; questions mismanagement within BOR, 49; and selection of Clarence McDonough as LCRA general manager, 46–48; visits LCRA damsites, 63
Buchanan Dam, *18, 19, 20, 20, 22*, 53, 64–65, 78–79; cracks in, 60–62; engineering problems of, 47; funding for, 59–60, 105; jurisdictional concerns over, 58, 60–61; partial completion of, 48, 52. *See also* Hamilton Dam
Bunger, Howard P., 50, 58, 60
Bureau of Reclamation: and assistance from Denver office, 52, 59–60; and fed-

Bureau of Reclamation (*cont.*)
  eral funding for flood control and irrigation, 44, 48–52; map of 1935 projects by, *32*; and Marshall Ford Dam, 66–67, 95, 111; promotion of flood control by, 41–42; questions about reputation of, 61–62; reaction of, to cracks in Buchanan Dam, 60–62; water resource development by, 26–27
Burlew, E. K., 58, 85
Burnet, Texas, 18, 25, 96
Burnet Chamber of Commerce, 57

Callahan divide, 4
Campbell, Price, 88
capital, location of Texas, 5
Carmody, John, 97–98
Central Texas Hydro-Electric Company, 19, 35
Chappell, Frank W., 45
Chickamauga Dam, 68
Cisco, Texas, 89
Civilian Conservation Corps, 42
Civil War, 8
Civil Works Administration (CWA), 29, 42
Clayton, Jim, 117n19
Colorado City, Texas, 4
Colorado Lights, 102
Colorado Navigation Company, 8
Colorado River (in western states), 14–15
Colorado River (Texas): agriculture along, 12–13, 37, 106; charting of, 4–5, 12; conservation legislation and, 13; early settlement along, 5–6, 21, 116n8; first government funding to improve and dam, 8–9; flood control along, 9, 27, 37, 39, 42, 48–49, 74–75; flooding and damage along, 8–11, 44–45, 44–46, 46, 61, 62, 69, 83, 85, 86–87, 93; and flood of 1938, 86–92; implication of hydroelectric power from, 42; and irrigation, 16, 39, 42, 48, 74; maps of, 6–7, *32*, *105*; naming of, 4–5, 8; navigation on, 5, 9, 10, 12–13; population growth along, 10–11, 21–22; rafts in, 8–10; recreation on, 42; roots of development of, xv; water power sites along, 5; watershed of, 3–4, 12, 37, 86, 90, *105*, 115n1, 128n75
Colorado River Authority, 35, 50. *See also* Lower Colorado River Authority

Colorado River Commission, 14
Colorado River Company, 20–21, 33, 35, 36, 37, 41, 43, 46
Colorado River Improvement Association, 12, 13, 16
Columbus, Texas, 4, 24, 87, 96
Comal Power Plant, 107
Concho River, 4
Connally, John B., 140n49
Connally, Tom, 72, 79, 81, 87
Conservation and the Gospel of Efficiency (Hays), xv
Cooke, Morris L., 96–97, 98
Corcoran, Tom, 72

Dallas Power and Light, 54
Dean, W. V., 37, 38
Debler, E. B., 41
De L'Isle, G., 4; map by, 6–7
Dern, George G., 54
Dick, W. G., 87
Duke Power Company et al. v. Greenwood County et al., 63–65
Duncan, A. J., 55

Eagle Lake, Texas, 87
Eilers, A. J., 77
Elliott, R. T., 48
Emergency Committee for Unemployment Relief, 28
Emery Peck and Rockwood Development Company, 16–17, 18, 19
Engelhard, Fritz, 72, 87

Fargo Engineering Company, 46
Federal Emergency Relief Act, 29, 42
Federal Emergency Relief Administration (FERA), 29–31, 33, 48, 50, 57, 69
Federal Land Bank, 27
Fegles Construction Company, 18, 46, 58
Ferguson, Miriam A. ("Ma"), 28, 30, 37–38
Fight for Conservation, The (Pinchot), xvi
Fry, Roy, 47, 55–57, 58, 64, 70

Garner, John Nance, 110
Garwood Irrigation Company, 16
General Flood Control Act of 1936, 15
Gloyna, Emmett, 118n29

Goldthwaite, Texas, 5
Grand Coulee Dam, 25, 44, 60, 82, 95, 104, 105
Granite Shoals Dam. *See* Alvin J. Wirtz Dam
Gray, Howard, 88
Guadalupe-Blanco River Authority, 24, 34
Guadalupe River, 5, 17, 24, 62, 72
Gulf States Utilities, 54

Hamilton, George W., 119n38
Hamilton Dam, 18, 19, 20–21, 58; engineering feasibility of, 41; funding for, 34–36; naming of, 119n38. *See also* Buchanan Dam
Harper, S. O., 60
Hays, Samuel, xv
Hetch Hetchy, 99
Hill Country, 4
Hoover, Herbert, 15, 27, 98
Hoover Dam, xvii, 15–16, 22, 26–27, 28, 31, 38, 43, 52, 105
Holleman, Tom, 17–18
Hopkins, Harry L., 29, 30–31, 33, 54, 122n29
Hopkins, Welly K., 17–18
Houston Power and Light, 54, 103
Hughes, Charles Evans, 53
Hughes, Sarah T., 37–38, 40, 111, 123n52
Hunt, Henry T., 34, 36, 38–39, 40–41, 48
hydroelectric power, 54, 90, *106*

Ickes, Harold: and Alvin J. Wirtz, 104; attitude of, toward reclamation efforts, 33; on economic benefits of Colorado River project, 42; and funding of Colorado River project, 39, 50–51, 54–55; and investigative study of Colorado River, 34–35, 40–41; and Marshall Ford, 82–83; and ownership of Austin Dam, 77; and scrutiny of LCRA management, 80; and selection of general manager of LCRA, 47; visits Texas, 68
Inks Dam, 57, 58–59, 60, *62*, 67, 105
Insull, Martin J., 16, 19–20, 37
irrigation companies, 40
irrigation projects, 26, 33, 37, 40–41, 63

Johnson, Adam Rankin, 9, 14, 18
Johnson, Adam R., Jr., 31

Johnson, Lyndon Baines, 70–73, *71, 73, 82, 102, 113*; and Alvin J. Wirtz, 70–71, 101, 103, 112; arranges to lower LCRA debt, 82–83; campaigns for U.S. Senate, 104–105; and FDR, 70–73, 98, 112; and Harold Ickes, 74–80; and James Roosevelt, 82; and presidency, 123n52; and REA funding for LCRA, 98–99; and renaming of Austin Dam, 95; as supporter of cheap electricity, 93–94, 100, 102–103; and Tom Miller, 75–77

King, Judson, 96
Kingsland, Texas, 5
Kubach, William, 51, 56

La Grange, Texas, 4, 87
Lamar, Mirabeau B., 5, 22
Lampasas Cut Plains, 4
La Sablonniere River, 4
La Salle, René-Robert Cavelier, Sieur de, 4
Lilienthal, David, 96, 100, 105, 126n41
Llano, Texas, 18, 55
Llano River, 4, 83
Lometa, Texas, 5
Lower Colorado River Authority (LCRA): Act, 37–40, 43, 70; advocates of, 31; bill to create, 35, 37–39; Board of Directors, 47, 48, 87, 89, 94, 124n19; and bond (debt) repayment by, 53, 93; David Lilienthal (TVA) visits, 105; and electrical power, 54, 89, 93–94, 96, 97–98, *106*; Elwood Mead visits, 51; and FDR, 52; federal funding for, 39, 40–42, 48–52, 56–57; and flood control, 26–27, 33, 39, 42, 74, 89–92, 107, 109, 111; and Harold Ickes, 40–41, 50–51, 68, 79–83; and investigation of 1938 flood, 88–92, 94–95; and irrigation, 33, 39, 41; James Buchanan visits, damsites, 63; labor strikes, 85; management of, 47–50; opposition to, 40, 53–59; and purchase of Buchanan Dam, 34–41; and role of public authority, 34–35, 40–41; and Rural Electrification Administration, 99, 106–108, 111; and water rights, 38; and "yardstick" concept, 99–100, 140n41
Lower Colorado River Flood Association, 129n82

INDEX   159

McDonough, Clarence, 47-48, 50, 52, 54-56, 58-59, 72, 75, 81, 87-88, 90, 112
McKenzie, A. J., 87, 88
McKenzie Construction Company, 67
Malott, C. G., 33, 36, 41, 43, 46-47
Mansfield, Joseph J., 36, 72, 79, 81, 87, 100, 110
Mansfield Dam, 100, 105. See also Marshall Ford Dam
Marble Falls, Texas, 5, 14
Marble Falls Dam, 49, 105
Marshall Ford Dam, 47, 49, 52, 59, 60, 65, 66-92, 84, 89, 105, 130n3
Matagorda Bay, 4, 8
Max Starcke Dam, 105
Maxwell site, 59. See Marshall Ford Dam
Mead, Elwood, 26-27, 34, 40, 49-51, 59, 125n33
Metcalf, Penrose, 38
Middle Rio Grande Conservancy District, 58
Middle West Utilities Company, 16, 19
Miller, Tom, 70, 75, 77, 95, 100, 111
Mims, Lewis, 55
Mississippi River, xvi, xvii; floods along, 14-16, 22
Mitchell, L. H., 34, 40-41
Moody, Dan, 90-92
Morgenthau, Henry, 54
Moritz, E. A., 85
Morrison, Ralph W., 20, 37, 38, 46-47
multipurpose water resource development: components of, prioritized, 85; failure of, 86; federal aid for development of, 26-27, 33-34; and hydroelectric power vs. flood control, 55, 91, 98-100; LCRA as most extensive, west of Mississippi River, xvii, 111; merits of, 13, 37, 81, 109; New Deal, xv, 107-108, 111
Muscle Shoals, xvi, 15, 38, 40, 99

National Defense Act of 1916, 53
National Resources Committee, 72
National Youth Administration (NYA), 70
Neches River, 5
New Deal, xv, 13; court cases to uphold projects in, 63-64; and electrical power, 96-98; FDR's response to, 25-

New Deal (cont.)
26, 28, 98; at local and state level, 52; and multipurpose river development, 107, 109-14; and 1937-38 economic crossroads, 80; and public works, 15; in Texas, 31-32; and the West, 111-12; and "yardstick" concept, 55-56, 99
Newlands Reclamation Act of 1902, 25, 33, 111, 120n6
New York Power Authority (NYPA), 98
Norris, George, xvi, 96, 99
Norris, John A., 33, 41, 55

Oakville Escarpment, 4
O'Daniel, W. Lee ("Pappy"), 105

Page, John, 57-58, 68, 72, 74, 82-83, 84-87
Parker Dam, 105
Pedernales Electric Cooperative, 101
Pedernales River, 4, 83, 90
Pinchot, Gifford, xvi, 96
Popple, Henry, 4
Possum Kingdom, 52, 80
Public Works Administration (PWA): and Austin Dam, 75-76, 111; and direct federal loans, 20-21, 35; and funding for state relief, 30-31, 33, 35, 38-39, 41-42; imposes restrictions on LCRA, 43, 54, 95; investigation of, projects in Texas, 34; and job creation, 29; role of, in selection of LCRA general manager, 47

railroads, 8, 13
R. aux Cannes ou Rio di San Marco o Colorado, 5. See also Colorado River (Texas)
Rayburn, Sam, 72, 110
Reclamation Era, The (magazine), 95
reclamation fund, concept of, 26
Reconstruction Finance Corporation (RFC), 20, 27, 28
Relief and Reconstruction Act, 28
Republic Portland Cement Company, 57
rice crops, 12, 14, 37
rice farming, 37
Rio de San Marco ou Colorado, 4. See also Colorado River (Texas)
Rio Grande (Rio del Norte), 4

Rio Rouge (Red River), 4
Rivers and Harbors Act of 1916, 12
Roosevelt, Franklin D., 25, 28, 34, 35, 37–38, 52, 70, 73, 96–97, 99
Roosevelt, James, 82
Roosevelt, Theodore, xvi
Rural Electrification Administration (REA), 97, 103, 107, 112; loans, 97

Sabine Pass, Texas, 29
Sabine River, 5
Saint Lawrence Seaway, xvii, 98
San Antonio River, 62
San Marcos, Texas, 70
San Saba, Texas, 4
San Saba River, 4, 37
Seguin, Texas, 23
Shasta Dam, 25, 104, 105
Sheppard, Morris, 72
Shirley Shoals, 9
Slattery, Harry, 104
Smithville, Texas, 25
Southwest Texas State Teachers College, 70
Starcke, Max, 94, 103, 114
Starcke Dam, 94
Sterling, Ross S., 27
Streiff, Abraham, 90
subsistence colonies, 30
Sulak, L. J., 87
Sulphur Springs Draw, 4

Taft, William Howard, xvi
Taylor, Thomas U., 10, 75
Technical Memorandum on the Colorado River Flood of July, 1938, 90
Teijas (Texas), 4
Tennessee River, xvi, 3, 13, 14, 16, 68, 98–99, 100, 111
Tennessee Valley Authority, xv, 36, 39, 40, 42, 96, 98–99, 100, 111
Tenth Congressional District, Texas, 25, 72, 95, 104, 112
Texas A&M, 72
Texas Centennial, 42
Texas Democratic Party, 104
Texas Electric Service, 54–55
"Texas' Little TVA," xvii, 100, 104, 107, 111, 140n49
Texas Power and Light Company, 40, 54–55, 75, 103

Texas Relief Commission, 28–29, 30, 31, 41, 120n13
Texas State Board of Water Engineers, 14, 33, 35, 41, 45, 90
Texas State Constitution, 13, 34
Texas Utilities, 54
Thompson, W. H., 54
Tom Miller Dam, 95, 105. *See also* Austin Dam
Trinity River, 5

unemployment, 27, 29, 30
U.S. Army Corps of Engineers, 8, 9, 10, 12–13, 14; and Mississippi River flood, 14–15
U.S. Department of Interior, 33, 34, 36, 65, 72, 81, 85, 94–95
U.S. Department of Treasury, 55, 57
U.S. Geological Survey, 10
U.S. Supreme Court, 53–55, 56, 63–64, 72, 80
Utah Construction Company, 67
utility companies, xvi, 37, 54–56, 62–63, 64, 88–91, 93–94, 97, 99, 100–101, 103

Valley Conservation and Reclamation District, 34

Walter, Raymond F., 51, 60, 62
Waterloo, Texas, 5
water mills, 5
water rights, 38
W. E. Callahan Construction Company, 67
Westbrook, Lawrence, 29–31
West Texas Chamber of Commerce, 16, 38
West Texas ranchers, 37–39, 111
West Texas Utilities Company, 88
Whipple, E. M., 68
Williams, Aubry W., 29–30
Wilson, Woodrow, xvi, 121n19
Wilson Dam, 16, 53, 98
Wirtz, Alvin J., 71, 82; as advocate for water development, 24–25; and Austin Dam, 80–81; and Brown and Root, 68; and Buchanan Dam reservoir, 94–95; as delegate to Democratic National Convention, 104; and Department of Interior, 104, 140n50; and funding of Marshall Ford Dam, 67–68; and Holle-

INDEX   161

Wirtz, Alvin J. (cont.)
man incident, 17–18; and James Buchanan, 66, 70; as legal counsel, 16, 55–58, 68; and Lyndon B. Johnson, 70–71, 85, 86, 94, 101; and opposition to private utilities, 55–57; and receivership of Hamilton Dam, 20–23; and Roy Fry, 56–57; and settlement received from Colorado River Company, 46; as sponsor of Texas projects, 33–35, 42; and support of C. G. Malott for LCRA general manager, 47; supports rural

Wirtz, Alvin J. (cont.)
electrification, 94–95, 101; testimony of, on July, 1938, flood, 91–92; visits Texas, 68–70; and Welly K. Hopkins, 17–18
WOAI, San Antonio, 103
Woodward, Walter C., 38
Work Projects Administration, 11

Young Owen D., 138n29
"yardstick" concept, 55–56, 99–100, 140n41

Damming the Colorado was composed into type on a Compugraphic digital phototypesetter in ten point Trump Medieval with two points of spacing between the lines. Friz Quadrata was selected for display. The book was designed by Jim Billingsley, typeset by Metricomp, Inc., printed offset by Thomson-Shore, Inc., and bound by John H. Dekker & Sons, Inc. The paper on which this book is printed carries acid-free characteristics for an effective life of at least three hundred years.

TEXAS A&M UNIVERSITY PRESS : COLLEGE STATION

www.ingramcontent.com/pod-product-compliance
Lightning Source LLC
Chambersburg PA
CBHW031248290426
44109CB00012B/488